AF689368

Shaping the Future of Work

Shaping the Future of Work

A Handbook for Action and a New Social Contract

Thomas A. Kochan

Lee Dyer

MITXP

© 2017 Massachusetts Institute of Technology

MITxPress imprint
Distributed by the MIT Press

All rights reserved. No part of this book may be reproduced in any form by any electronic or mechanical means (including photocopying, recording, or information storage and retrieval) without permission in writing from the publisher.

ISBN: 978-0-262-53491-8

Dedication

To the current and future generations of workers and their families

Contents

Preface

This book is part of a larger project aimed at changing the course the American economy and especially our employment relations system has been on for the past 30 years. Since the 1980s, the economy has been reasonably productive and has worked reasonably well for investors and high-level executives. But it has not worked so well for ordinary working Americans and their families. During the same period, many of the institutions that support workers' rights atrophied and lost their ability to build an inclusive economy in which the fruits of prosperity are shared widely and equitably.

Concern over these long-term trends motivated us to create MBA and online courses at MIT and a companion course at Cornell University's ILR School. We have taught these courses for two years and intend to do so in the future. Given recent developments in the world of work and the larger environment in which work is situated, we felt the need to update much of our material and to focus more directly on the need for leaders from business, the work force, education, labor, and the government to come together to build what we call *a new social contract* to make work pay off better in the years ahead for all concerned parties.

The most obvious big new development in the environment is the deep division that was laid bare by the 2016 presidential election and by parallel developments around the world. This division is between those who have done well and see the future of work as full of opportunities and those who feel that they and their families are being left behind and are angry about their current circumstances and worried about their futures. Brexit, the election of Donald Trump as president of the United States, the growing popularity of extreme political groups and candidates across Europe, the backlash against immigrants, concerns over terrorist attacks on citizens, and worries that future technologies will eliminate more jobs than they will create have put the world on edge. These tensions have added a sense of urgency to our task. Unless we begin to heal these divides and address these worries, our economy and our society could descend into a state of disarray where ultimately everyone loses.

We think there is an alternative path to the future and we believe that path has to start by making work meet the needs of all those who depend on it: workers and their families, employers and the larger business community, and society. The path we propose does not depend on stoking current or future levels of anger or frustration, nor do we have any interest in trying to turn back the clock to the economy and working world that functioned so well in the post–World War II period. Instead, we challenge each major stakeholder group to *change* in some key ways to get the economy moving in a way that creates good jobs with adequate wages for all. This can be viewed as an alternative narrative to the rhetoric of the extremists of the left and right. They are certainly correct to call for change, but these changes must be grounded in evidence, experience, and innovations that produce broadly shared benefits.

What follows is more than just a blueprint of what needs to be done; we also suggest practical steps the principle stakeholders can take. We emphasize three key points. First, all business leaders make choices about how they compete, and the choices they make have predictable consequences for the quality of jobs that are created and sustained. We urge them to choose *high-road business and employment strategies* that are capable of generating both good financial returns and good jobs. Second, we encourage workers to be proactive in shaping their future employment opportunities and careers. We encourage workers to invest in *lifelong learning* to ensure that they obtain and maintain the skills they need to contribute to a knowledge- and innovation-based economy. And we encourage workers to band together to amass *new sources of power* to fill the current void in bargaining power in the labor market. From these new bases, they will be able to demand middle-class wages (or better). Third, no individual stakeholder group—whether it be business, workers, educators, labor groups, or government policy makers—can successfully turn the economy around on its own. We call for *collaboration* among these groups so they can overcome and solve the market and institutional failures that are holding back progress today. By doing so, these stakeholders can forge a new social contract that is capable of generating and sustaining an economy and a labor market that works for all.

This book focuses on actions that are needed in the United States. While we are aware that many workers and societies around the world are facing the same or similar challenges, we discuss what needs to be done here, where most of the research and experiences we draw on are situated. Perhaps this book will motivate people from around the world to discuss how these U.S.-based ideas for shaping the future of work can be imported and adapted to fit their economies, cultures, and institutions.

Notwithstanding the current worries, we are optimistic about the possibilities for the future, in large part because we have seen in our own research and in our involvement with these issues how individual and collective actions can make a difference and can produce great results for business and nonprofit organizations while also creating great jobs and careers for those who help achieve these results. This is why we believe that we are not just pawns controlled by globalization, technological changes, or any other force totally outside our control. If the key stakeholders take the right actions and work together, they can shape the future of work to everyone's benefit.

This book is one component of a broader array of efforts to chart a path toward a new, more equitable social contract. At MIT, it is embedded in a major new research, teaching, and outreach initiative called Good Companies–Good Jobs that focuses on increasing the number of organizations and work relationships that do just what the title implies: produce good financial returns for shareholders and good jobs and careers for workers. On our website (http://iwer.mit.edu/speak-up-for-work) you can sample the videos we have prepared for the online course we refer to in the chapters and you can participate in a social contract negotiations exercise so you can compare your expectations and proposals for the future with those of our students and of others who have completed it. At Cornell, the ILR School is dedicated to "Advancing the World of Work" in part by educating current and future labor leaders, human resource professionals, and government policy makers and in part by conducting research and outreach activities through a number of institutes and centers, including the Worker Institute (http://www.ilr.cornell.edu/worker-institute) and the Center for Advanced Human Resource Studies (http://cahrs.ilr.cornell.edu/Home.aspx). We want to bring other voices into the conversation as well and in doing so expand the circle of those who are committed to paving a positive path toward a new and more viable social contract for work.

Because the world of work is constantly evolving, we see this book as a living document. We have use some material from an initial version that was published in January 2016 and have introduced new material that reflects our current thinking. We intend to continue to follow this pattern in the years ahead. Fortunately, our publisher, MIT Press, has launched a new venture, MITxPress, to support this type of publication. We are pleased to be part of this new initiative. Between editions we will also post updated comments, new materials and data, and inputs from our students on the MIT website (http://iwer.mit.edu/speak-up-for-work).

Finally, a word about what brings the two of us to work together. We have been close friends and colleagues since our graduate school days at the University of Wisconsin. Our families have grown up together, and we have solved world problems (in our heads, at least) during long walks and while doing dishes on the Thanksgiving holidays our families have celebrated together for the last 40-plus years. But never before have we been more worried about the legacy our generation is leaving to the next. We want a better future, not only for our own children and grandchildren but also for all their peers of all races, genders, countries of origin, and socioeconomic status. We feel a sense of urgency and hope that others do as well and will join us in making the future of work better for all who depend on it now and will be depending on it in the years to come.

Thomas A. Kochan, Cambridge, Massachusetts

Lee Dyer, Ithaca, New York

Acknowledgments

Writing this book and teaching the online and on-campus courses that are linked to it required help and coordination from a small army of talented colleagues. We are fortunate to be surrounded by a host of creative people who helped bring this book together in record time and in good humor.

We thank members of our creative MITx production team who helped integrate the text with the videos and other online materials: Kenny Caudhill, Brad Kay Goodman, Lana Scott, Shelley Upton, Geoff Wilson, and Mary Ziegler. Alex Kowalski, Grey Sullivan McCune, Emily Wang, and Maya Kolaczynski provided TA support and made sure everything came together in a seamless fashion. Jacqueline McGoldrick provided administrative support and put her creative touches to work in preparing and improving some of the more complicated figures and charts scattered across the different chapters. John McCarthy, Christine Riordan, and Joel Cutcher Gershenfeld updated videos from prior classes. Mike Haberman revised and reprogrammed the Social Contract Exercise.

At Cornell, superb administrative assistance was provided by Pam Staub and Jo Hagin with great tolerance for tight deadlines and quick turn-around times.

We very much appreciate the financial support for this work provided by the MIT Good Companies-Good Jobs Initiative, the Mary Rowe Fund for Conflict Management, the MIT Sloan School of Management, the MITx Office of Digital Learning, and Cornell's Center for Advanced Human Resource Studies (CAHRS).

One of our key goals for this project was to produce and publish this book in a timely fashion. We are grateful for MIT Press and its Economics, Finance, and Business Editor Emily Taber for suggesting they publish the book in their new MITxPress collection. We are delighted that MIT Press has created this new publication option and are pleased to be one of the initial offerings to come through this channel.

We are indebted to many colleagues at MIT and Cornell who helped inform and shape our thinking about the issues covered in the book. At MIT faculty and students in the Institute for Work and Employment Research are an on-going source of inspiration and new research. The same is true of the faculty and students at Cornell's ILR School, and especially those affiliated with the Department of Human Resource Studies and CAHRS.

Two colleagues and friends served as our expert sounding boards and advisors on various ideas and technical points throughout this project. Barbara Dyer has been a partner in teaching these materials at MIT Sloan and helped sharpen many of the ideas covered in the book. We are pleased that Barbara is now leading our new Good Companies-Good Jobs Initiative at MIT. Wilma Liebman clarified and corrected key points related to labor law and innovations in worker voice and representation and did so in her usual quiet and competent manner.

It is not always the case that authors finish a project with warm feeling toward the editor who has the audacity to alter their prose. In this case, we are fortunate to have turned our draft chapters over to Kate Babbitt, the best editor we have ever worked with. Kate not only clarified and improved the written words, she developed and assembled a loose collection of chapters into a final product that could go directly to print. Her knowledge of these issues and her passion for improving the world of work made it all the more fun to debate, and then almost always accept, her edits, requests for clarification, and gentle nudges to write for normal people rather than our insider academic crowd.

Finally, we thank our families, not only for putting up with the time we spent on this project, but for the many years of shop talk they have endured during our times together. We can't say we convinced them all of the wisdom of our ideas, but in this case tolerance is all we have needed and undoubtedly will need as we subject them to more of the same in times to come.

Figures

Tables

1

America's Challenge and Opportunity: Building a New Social Contract at Work

The 2016 U.S. presidential campaign and election, the UK vote to exit the European Union, and growing support for more extreme candidates in several European countries have exposed the deep-seated frustrations, anger, and worries of those who feel left behind by recent economic forces. In the United States, the pressures have been building for several decades, fueled by rising income inequality, flat wages, the loss of good jobs and job opportunities for people living in regions most deeply affected by globalization, and worries that technology will eliminate more jobs in the future. For years, a majority of Americans have said the country is going in the wrong direction and that they fear the next generation will have a lower standard of living than they experienced growing up.[1]

Yet the anger and response came as a surprise to many who see the future through very different eyes. For example, both of the authors of this book work in great universities where new scientific discoveries, technological breakthroughs, and organizational innovations are constantly being announced and turned into new products, services, and opportunities. We interact with business leaders who are eager to enhance their global competitiveness and hungry to attract the scarce talent needed to make this happen. And we teach students who are eager to put new technologies and opportunities to work to solve the world's big problems and who expect to build successful, rewarding careers. From these perspectives, it is easy to focus more on future prospects than on current problems.

These two world views emanate from observations of the same set of forces: globalization and advancing technology. But are these forces outside our control? Are either the pessimists or the optimists destined to be right? We don't think so, unless, of course, we do nothing to address the underlying challenges or, worse, make a futile attempt to turn back the clock to an earlier and simpler time. Instead, we propose a different path, one in which all sectors of society are mobilized to work together to create competitive enterprises that combine high-quality jobs with rising wages and

[1] Elizabeth Mendes, "In U.S., Optimism for Youth Reaches an All-Time Low," *Gallup*.com, May 2, 2011, http://www.gallup.com/poll/147350/optimism-future- youth-reaches-time-low.aspx.

opportunities, while also healing the wounds that divide us. In what follows, we will outline the steps that are needed to build what we call a new social contract at work, one that is based on mutual respect and is attuned to the needs of today's and tomorrow's economy and work force.

Here's the basic argument of the root causes of today's divide and the strategy we propose:

The main source of the challenges the work forces of today and tomorrow face is the fact that the rapid pace of globalization and changes in technology and demographics have outpaced many of the public policies, business strategies, and organizational practices that were designed in an earlier era to govern work, pay, and employment relations. Closing this gap by updating these policies, strategies, and practices is essential to building a world of work where all can prosper.

Preview: Policies and Business Models to Support Great Companies and Great Jobs

To whet your appetite for this argument, let's illustrate two things: first, how outdated our employment policies are; and second, how the models that guide the current business strategies of corporations need to change.

Most of our labor and employment legislation dates back to the New Deal of the 1930s. That flurry of action was a direct response to the Great Depression and a belated response to the shift from farming to an industrial economy. Not surprisingly, given work and family patterns at that time, the framers of this legislation and the workplace practices that followed had a model of the typical worker in mind. That worker was a male production employee who worked full time under close managerial supervision in a large domestic firm. Conveniently, he had a wife at home to attend to family and community responsibilities.

Today, in contrast, we have an economy that is knowledge driven and that values innovation. The work force is diverse. Nearly as many women as men are working in the paid labor force. People can expect to move across employers multiple times in their careers and in and out of full-time and part-time work so they can attend school and/or take care of family responsibilities. It is not even always clear who the employer is, given the advent of franchise, contractor, and outsourcing arrangements. Yet most of our laws and the regulations and procedures used to enforce them still reflect the earlier era. For example, the United States is the only large industrial economy that still lacks a national policy on paid family leave. This is just one in a range of policy innovations and/or updates that we need, not just to catch up with the rest of the world but also to better meet the needs of our knowledge-driven economy and diverse labor force.

The example that illustrates how U.S. business models need to change comes from a courageous group of employees at Market Basket, a New England grocery chain with 71 stores and 25,000 employees. Their actions highlighted a debate that needs to be raised all across America: namely, what is the purpose of a business? Is it just to make money for shareholders or is it to make money *and* provide good jobs for employees and good service and fair prices for customers?

For six weeks in the summer of 2014, executives, store managers, clerks, truck drivers, and warehouse employees of this family-owned business stood side by side outside their stores demanding that their CEO be reinstated and that the business model that made the company thrive and supported good jobs, low prices, and great customer service be maintained. Their customer base cheered them even though they had to shop elsewhere at considerable inconvenience and expense. Never before have we seen such a broad coalition of workers and customers unite to save a business from short-sighted shareowners looking to extract more cash to line their pockets. But they did so at considerable risk, because the protesting employees who were managers and supervisors had no legal protections under our outmoded labor law, the Wagner Act, which dates back to 1935.

Yet they persevered. The combination of the pressure this coalition created, a tremendous outpouring of community support, and creative use of social media to maintain solidarity pushed the board of directors to relent and sell the company to the beloved CEO, who brought back the work force and the store's customers and restored community goodwill.

We will build on this case at various points in this book because it illustrates both the frustrations many in society experience about what is wrong in American business and a positive way these frustrations can be turned into collaborative actions that create change. The reason the Market Basket employees gained such broad and deep public support is that they were seen as fighting to preserve what we and others call a "high-road" business strategy and set of workplace practices that can deliver good profits to shareholders, good jobs and careers to employees, and good prices and service to customers (see Fig. 1.1). This is what the American public wants to see in business and at work today and in the future, and it is why the public, customers, and employees all rallied together to keep Market Basket from sliding from the high-road to the low-road strategy. Our challenge is to make the high-road model for business strategies and employment relations the norm, not the exception to the more dominant approach, which treats employees as just another cost to be minimized, tightly controlled, and disposed of when not needed.

Figure 1.1. Effects of High-Road and Low-Road Company Strategies

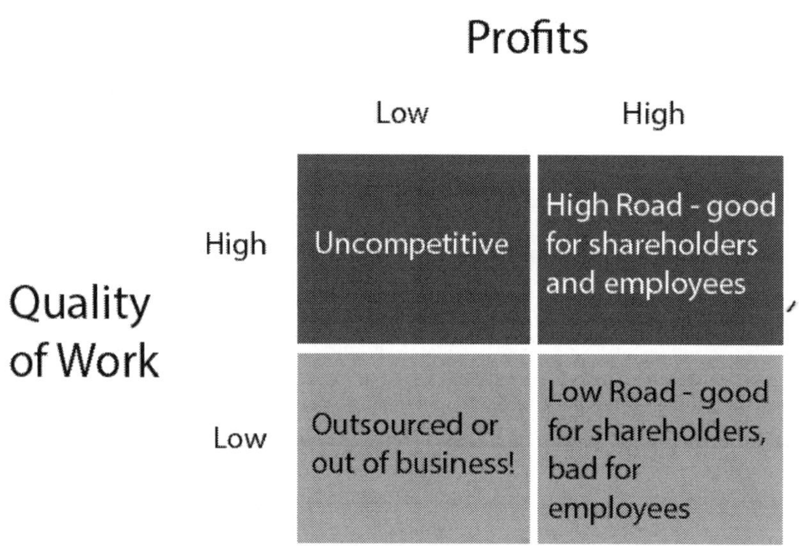

So let's get started, first by painting a quick picture of the challenges and opportunities today's workers face, with a special eye for young people who are just entering the labor force.

A Two-Dimensional Jobs Crisis

The first decade of the twenty-first century earned the sad title of the "lost decade." Workers of almost all occupational and income levels were the losers in two dimensions: the quantity and quality of jobs. The second decade is struggling to become known as the recovering decade. Figures 1.2 and 1.3 tell these stories.

Figure 1.2 tracks how many jobs were lost during the Great Recession of 2007-2009 and how many still need to be created to recoup them and to keep up with the growth in the labor force since then. Job growth was very slow from 2009 to 2013 and then began to pick up in recent years, averaging nearly 200,000 jobs per month. *Yet a decade after the beginning of the recession, the economy still has not generated enough jobs to make up for jobs that were lost and to absorb the number of young people who have entered the labor force since then. Even worse, the quality of jobs being created is, on average, lower than those lost in the recession.* This is the worst job creation record of any recession since the Great Depression of the 1930s. Let's hope 2014–2016 are the models for the future, not 2009–2013.

Figure 1.2. Jobs Gap Since the Great Recession of 2007–2009

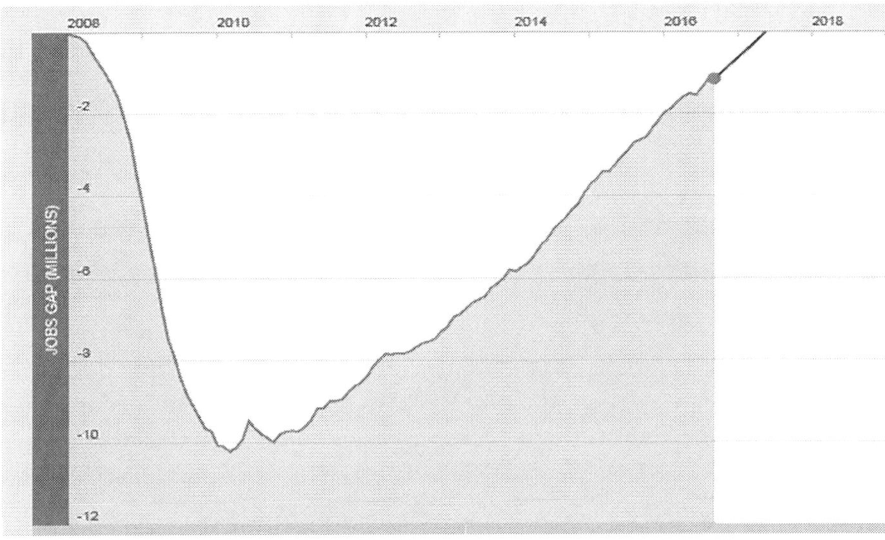

Source: Hamilton Project Jobs Gap Interactive Tracker, January 2008–September 2016. http://www.hamiltonproject.org/charts/closing_the_jobs_gap.

How does this affect young people entering the labor market? As late as 2014, 46 percent of college graduates were not finding jobs that would allow them to put to work the knowledge and skills they learned in college.[2] They are what we call underemployed—they are working in low-wage retail, restaurant, or other service jobs that don't require a college degree. These jobs don't put their skills to work or provide opportunities for further learning and development, and they likely pay wages that are hardly enough to enable young workers to meet their college debt payments, much less start a career and/or a family. To make things worse, a significant body of

[2] Jaison R. Abel and Richard Deitz, "Are the Job Prospects of Recent College Graduates Improving?" *Liberty Street Economics*, September 4, 2014, http://libertystreeteconomics.newyorkfed.org/2014/09/are-the-job-prospects-of-recent-college-graduates-improving.html#.VUt7TvlVhBc.

research indicates that the imprint of starting a career in this type of depressed labor market lasts for a long time, in some cases for one's entire working career. Not finding a career position with a decent starting wage and opportunities for continued learning and advancement imposes significant and in some cases permanent damage.

Figure 1.3. The Social Contract, 1947–2013

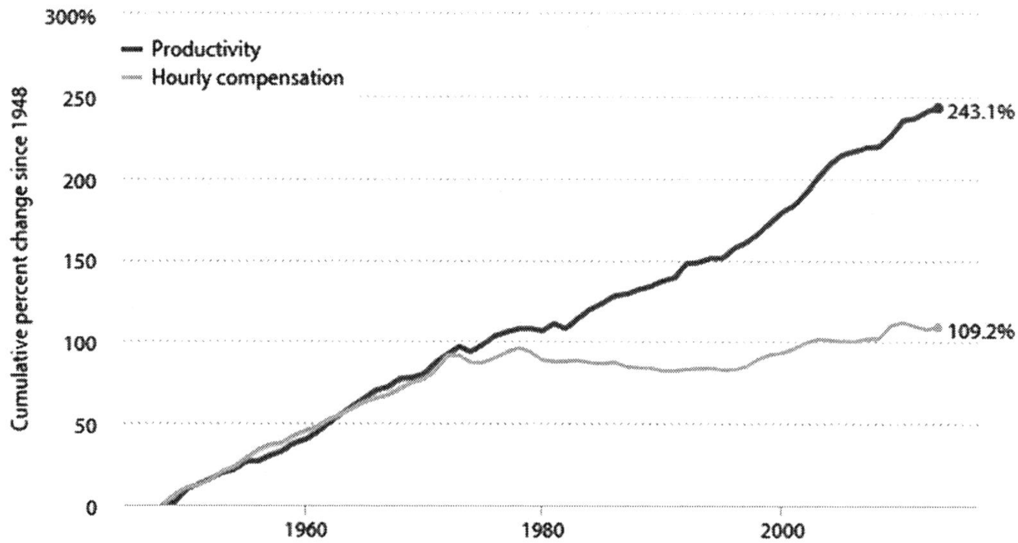

Source: Josh Bivens, Elise Gould, Lawrence Mishel, and Heidi Shierholz, "Raising America's Pay: Why It's Our Central Economic and Policy Challenge," *Economic Policy Institute*, June 4, 2014, http://www.epi.org/publication/raising-americas-pay/.

Young students in our online course provided information about their first-hand experiences with the structure of the current U.S. labor market. One coined a name for it: "working nomads."

> I think the concept of working is dramatically changing in my generation ([born in the] 80s and younger), and the change couldn't [be] understood by [an] older generation. We want to work at a stable organization, but [those] jobs are vanishing so have to work as an unpaid intern or part-time worker. . . . "Working nomads" are a growing tendency of today's working trends, I think. . . . So does this trend entirely change our job structure or [is it] just a temporary trend? I'm not sure, but we should focus on this tendency [in order] to understand our generation and today's world.

There is another group that has been very badly hurt by the recent job trends—those who have lost good-paying jobs in communities that have been hardest hit by globalization and technological changes in manufacturing industries. The United States has lost one-third of its manufacturing jobs since the 1970s. The communities that were hit hardest by this decline have experienced long-lasting drops in income, and a large number of workers affected have experienced long-term unemployment. Some have dropped out the labor force altogether. Those who did find new jobs have experienced on average decrease in wages of 15 to 20 percent.

Figure 1.3 tells the story of the second dimension of the jobs crisis. Look at the 30 years since about 1980, a time when earnings have essentially flatlined. Over the course of those years, the productivity of American workers grew by a healthy 80 percent, but family income grew by about only 10 percent and average hourly wages inched up only about 6 percent. The data in Table 1.1 indicate why the first decade of this century earned its "lost decade" label. Real compensation (wages and benefits adjusted for increases in the cost of living) either declined or did not increase for high school or college graduates. Until 2015, only those at the top of the occupational ladder with advanced degrees experienced modest wage growth. 2015 once again showed promise on the income front; for the first time in this decade incomes grew on average by 5.2 percent, including incomes for those at and near the bottom of the income distribution.

Table 1.1. Total Changes in Real Earnings, 2000–2011

Group	Employment share (%)	Earnings change (%)
Less than high school	8.4	–12.5
High school graduate	27.4	–4.1
Some college	27.8	–8.9
College graduate	23.2	–8.5
Masters' degree	9.5	–3.7
MD, JD, or MBA	1.9	+2.2
PhD	1.8	+3.4

Source: Jonathan Haskel, Robert Z. Lawrence, Edward E. Leamer, and Matthew J. Slaughter, "Globalization and U.S. Wages: Modifying Classic Theory to Explain Recent Facts," *Journal of Economic Perspectives* 26, no. 2 (2012): 119–140.

Where have all the fruits of increased productivity gone in the last 30 years? Figure 1.4 tells this story. Most of the income growth went to the top 1 percent or less of the population. When members of the Occupy movement claimed in 2012 that too much of the nation's income was concentrated in the top 1 percent of the population, they had their facts right. America is now suffering from the highest level of income inequality of any time since the 1920s. Figure 1.5 provides an even more vivid picture of how the shares of national income have shifted, Since the 1980s, the share going to the top 1 percent doubled from 10 to 20 percent, while the share of the national income in the hands of the bottom 50 percent declined from 19 to 12 percent.

Figure 1.4. Annual Income Share of the Top 1 Percent of the U.S. Population, 1910–2010

Source: Emmanuel Saez, "Striking It Richer: The Evolution of Top Incomes in the United States," updated with 2009 and 2010 estimates, March 2, 2012, http://elsa.berkeley.edu/~saez/saez-UStopincomes-2010.pdf.

Little wonder that this became a pivotal issue in the 2016 presidential election campaign. While candidates differed on what to do about it, none could ignore it because voters voiced their frustrations—even anger—loud and clear. The message: The economy is not working for us and we are not going to take it much longer. Politicians from across the political spectrum, business leaders from Warren Buffett to Jamie Dimon (see Figure 1.6), and even Pope Francis expressed the worry that if this divide persists, it will do more than just limit economic growth: it could also threaten the future of our democracy, just as extreme inequality has done in other countries in years past.

Figure 1.5. Shifting Shares of Pre-Tax National Income, 1980–2014

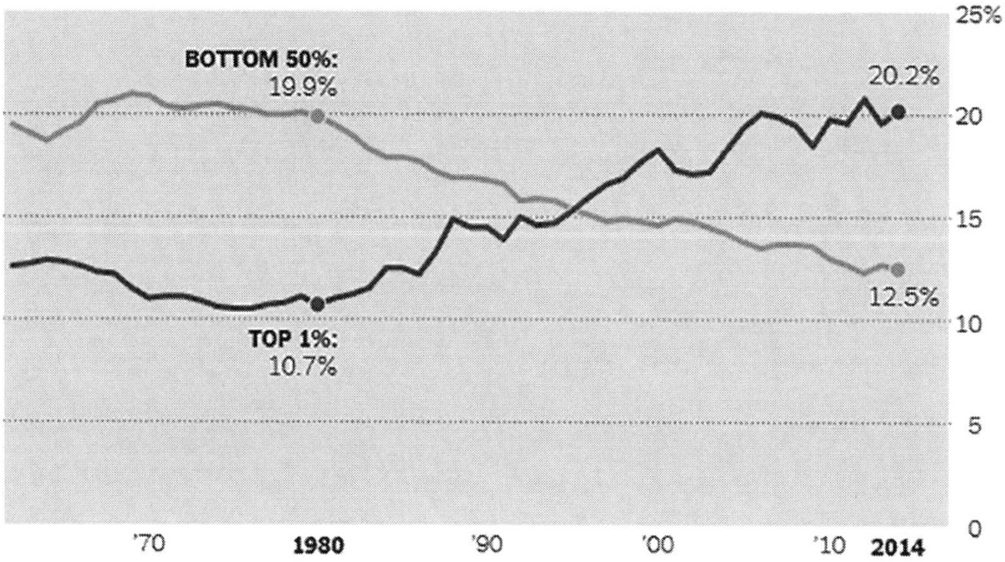

Source: Data from Thomas Piketty, Emmanuel Saez, and Gabriel Zucman reported in "Patricia Cohen, "A Bigger Economic Pie but a Smaller Slice for Half of the U.S.," *New York Times*, December 6, 2016, http://www.nytimes.com/2016/12/06/business/economy/a-bigger-economic-pie-but-a-smaller-slice-for-half-of-the-us.html?_r=0&imm_mid=0eb6e9&cmp=em-business-na-na-newsltr_econ_20161209.

A glance at the earlier years covered in Figure 1.3 suggests that wages and incomes didn't always lag behind growth in productivity. Indeed, from the mid-1940s through most of the 1970s, these two economic indicators moved pretty much in tandem. This fact will feature prominently in our discussion throughout these chapters. The tandem movement of wages and productivity over the three decades following the end of World War II captures the essence of what we call the post–World War II social contract at work. Clearly this contract broke down in the 1980s and has remained broken ever since. We will have to figure out what the next generation's social contract could look like.

Figure 1.6. Comments on Income Inequality

Pope Francis	"I ask you to ensure that humanity is served by wealth and not ruled by it. The growth of equality demands something more than economic growth, even though it presupposes it. It demands first of all a transcendent vision of the person. . . . It also calls for decisions, mechanisms and processes directed to a better distribution of wealth, the creation of sources of employment and an integral promotion of the poor which goes beyond a simple welfare mentality." "Pope Francis' Message to World Economic Forum in Davos," *News.Va*, January 21, 2014, http://www.news.va/en/news/pope-francis-message-to-world- economic-forum-in-da.
President Barack Obama	"We know that people's frustrations run deeper than these most recent political battles. Their frustration is rooted in their own daily battles—to make ends meet, to pay for college, buy a home, save for retirement. It's rooted in the nagging sense that no matter how hard they work, the deck is stacked against them. And it's rooted in the fear that their kids won't be better off than they were. And that is a dangerous and growing inequality and lack of upward mobility that has jeopardized middle-class America's basic bargain—that if you work hard, you have a chance to get ahead." "Remarks by the President on Economic Mobility," White House website, December 4, 2013, http://www.whitehouse.gov/the-press-office/2013/12/04/remarks-president-economic-mobility.
Jamie Dimon, CEO, JP Morgan Chase	"Wage stagnation. Income inequality. . . . Issues like these have led approximately two-thirds of Americans to believe that the next generation will be worse off than the last. And it is true that too many people are not getting a fair opportunity to get ahead. We must find ways to help them move up the economic ladder, and everyone—business, government and nonprofits—needs to play a role." Jamie Dimon, "Why We're Giving Our Employees a Raise," *New York Times*, July 12, 2016, http://www.nytimes.com/2016/07/12/opinion/jamie-dimon-why-were-giving-our-employees-a-raise.html?_r=0.
CEO and venture capitalist	"We've had it backward for the last 30 years. . . . Rich businesspeople like me don't create jobs. Rather they are a consequence of an ecosystemic feedback loop animated by middle-class consumers, and when they

Nick Hanauer	thrive, businesses grow and hire, and owners profit. That's why taxing the rich to pay for investments that benefit all is a great deal for both the middle class and the rich." "Too Hot for TED: Income Inequality," *National Journal*, May 16, 2012.
Warren Buffett, CEO, Berkshire Hathaway	"The Forbes 400, the wealthiest individuals in America, hit a new group record for wealth this year: $1.7 trillion. That's more than five times the $300 billion total in 1992. My gang has been leaving the middle class in the dust." "A Minimum Tax for the Wealthy," *New York Times*, November 25, 2012, http://www.nytimes.com/2012/11/26/opinion/buffett-a-minimum-tax-for-the-wealthy.html.

Trends in job satisfaction also track this breakdown in the social contract. Figure 1.7 uses data from the Conference Board, a highly respected business research organization, to track national trends. For the first time since the Conference Board began collecting these data in the 1980s, less than 50 percent of the work force report that they are satisfied with their jobs. Note that young workers report the biggest decline in and the lowest level of satisfaction. In 2013, only 28 percent of workers under age 25 said they were satisfied with their jobs. That number recovered from its low point but was still below 50 percent in 2016.

Figure 1.7. National Trends in Job Satisfaction, 1987–2015

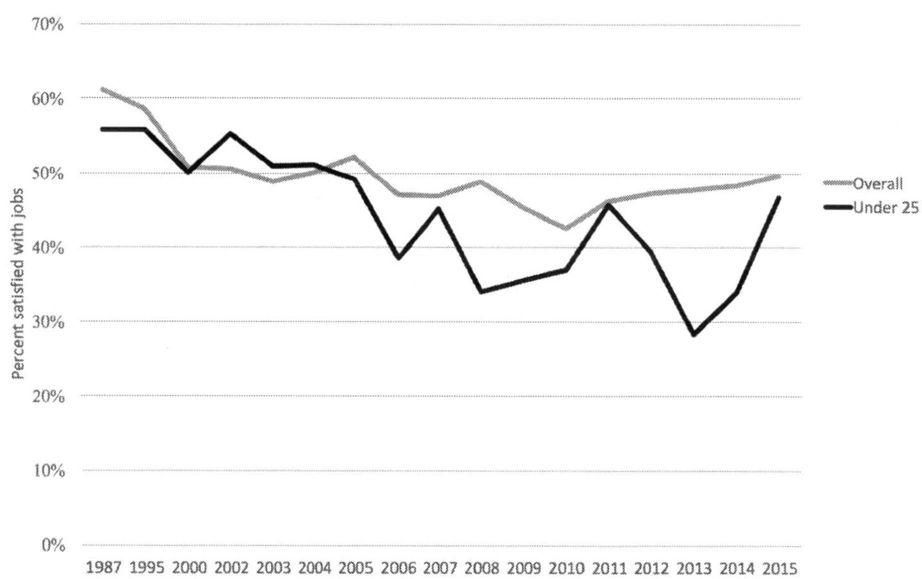

Source: The Conference Board.

Taken together, these data paint a challenging picture. They tell us that economic conditions play a strong role in the ability (or willingness) of employers to create jobs. These same forces, along with other factors discussed later, affect the quality of the jobs firms create and sustain. But employers don't operate in a vacuum. Rather, they operate in a broader ecosystem of potentially powerful institutions—the government, labor unions/labor consortia, societal norms and expectations, educational systems, and so on—that also influence to some extent the number of jobs and to a large extent the quality of jobs that are created and the quality of the support system for those who lose their jobs. In the postwar period, these forces were such that the American Dream was intact for most people. Since the 1980s, however, things have taken a turn for the worse in terms of wages and other conditions that affect job quality, and since the Great Recession the challenges have extended to a shortage of jobs. Employers are squeezed, unions have lost much of their clout, and government policies and educational systems haven't kept pace with changes in the economy or the nature of work. As a result, it has become difficult to realize the American Dream today, and it promises to get much more difficult for the generations to come unless something is done to change the path we're on.

The Political Dimension: Is Washington Asleep or in Gridlock?

If a modern-day Rip Van Winkle were to wake up today after a 30-year nap, she or he would very likely look at these facts and ask "Why have the leaders of our economic and political institutions let our economy and society slide for so long? Were they all asleep with me?"

And I suppose we would respond: "Good question, but no they were not totally asleep. They have been stuck in ideological gridlock for at least the last decade and in some cases, in particular the relationships between business and labor leaders, for even longer. The sad reality is the country is more divided politically today than at any time perhaps since the 1930s."

Rip's equivalent might then ask, "But what about now? I'm hearing that the new president is going to 'make America great again.' Isn't that a hopeful sign that things will change?"

Our answer to this would have to be "We don't know. It is too early to tell." We hope so, but we also hope that leaders in Washington take time to learn about all the innovation that is already under way at local and state government levels and in the private sector and civil society. History tells us that these local innovations can serve as "laboratories for democracy," as Supreme Court Justice Louis Brandeis said many years ago. So we will draw heavily on local and private sector innovations under way across the country to illustrate a potential positive and forward-looking path.

But we would also have to say there is a big hurdle to overcome. As we have already noted, the consequences of this 30-year gridlock came home to roost in the 2016 elections. America is deeply divided on a host of economic and social issues and there is little respectful dialogue between people who disagree or have different interests. So we have to invent a process for restoring respectful dialogue and search for common ground. That is why we use the metaphor of building a new social contract in this book.

The idea of a social contract was first developed by Jean-Jacques Rousseau and other philosophers to describe the ideal relationship between citizens and their government. We borrow it here to explore what constitutes a social contract at work. In doing so, we ask what are *the mutual expectations and obligations that workers, employers, and their communities and societies have regarding work and employment relationships*. To turn this question from an academic discussion into action, the parties that share an interest in and responsibility for shaping the expectations and obligations that govern work must engage with each other and search for norms and practices that are acceptable to all and that in the end help all parties realize its interests, goals, and expectations. This in turn means that we must restore dialogue among the key groups and foster a negotiation of sorts to search for common ground wherever that is possible and to respect the need for compromise when it is necessary.

In the chapters that follow, we will outline what we think has been learned from our history, the challenges that current economic realities pose, the opportunities that might lie ahead, and the innovative efforts that have emerged to take advantage of some of these opportunities. Our goal is to build on what we have learned from our research and experience over the years so that whatever dialogue and negotiations take place are informed by what we know or think we know about what works and what doesn't. This information is essential if we are to move forward. Listening and talking to only our narrow circle of like-minded people or favorite media outlets that say what we already believe to be true is not enough.

Destiny or Opportunity?

We have covered the bad news, the jobs crisis for young workers and more mature workers who have lost good jobs, and people's lack of faith in the leaders and institutions that govern work and employment. But with crisis comes opportunity. Indeed, one bit of good news is that the chorus of diverse voices noted in Figure 1.5 expressing concerns about income inequality suggests that we may be at a point where people, like the courageous Market Basket employees and customers, are ready to do something to address these issues. In fact, over the course of 2015 and 2016 a few glimmers of hope emerged as companies such as Walmart, McDonald's, and J. P. Morgan Chase announced their intention to raise wages above the required minimum. Regardless of whether one attributes their actions to a somewhat tightening labor market or a response to pressures from strikes of fast-food workers and national protests, these are good first steps in the right direction. Let's hope there will be more to come.

A second reason for hope comes from the motivations and aspirations of young people, reinforced by the views many students have expressed in an online class we taught at MIT. While these students may be disillusioned with American politicians, many are still positive about their own futures and are motivated to address the big problems they and others face at work, in their families, and in society. Take a look, for example, at how students ranked their goals and aspirations for future jobs.

Figure 1.8. Students' Aspirations Regarding Work

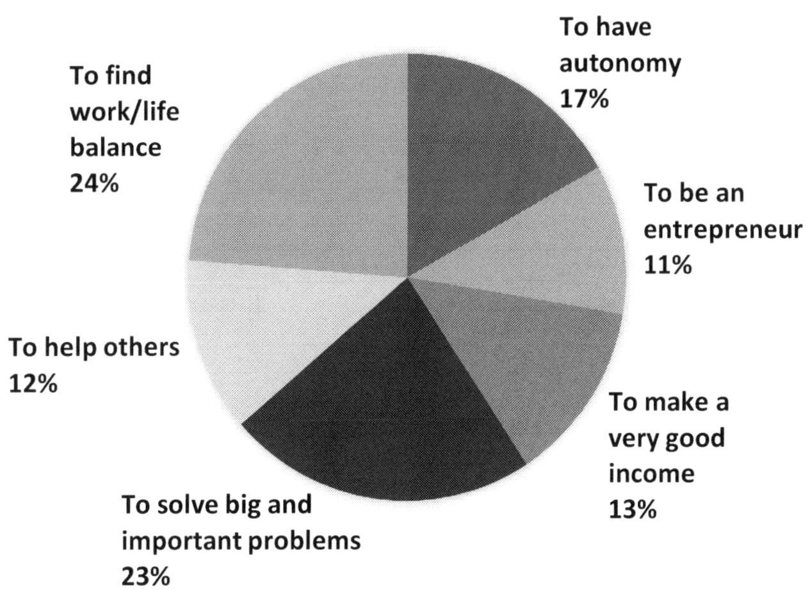

Source: Student responses to survey administered through the online course "The American Dream for the Next Generation," MIT Sloan School of Management.

At the start of the course, we asked participants to rank order their aspirations/priorities for work. The data in Figure 1.7 show that participants want to address big problems at work and at the same time have a sensible work-life balance. While doing well financially is also important, these data, like the findings of other surveys of both young and more experienced workers, show it is not at the top of the list.

The wide distribution of responses suggests that no single aspiration dominates. The comments students made reinforced this view. This extended quote from a young woman illustrates how she (and she seems to speak for others) sees these different aspirations fitting together:

> My dream includes a job where I do important work (working in health care data analytics)—something that gives back to the community. . . . I dream of having a work/life balance that melds together. . . . I hope that I'm able to utilize technology to make work more accessible and eventually have the option to work from home. I do hope to be able to find a company that has a retirement plan where the company contributes some percentage to the plan. In some regards, I hope to have some of the same benefits that my parents had—health insurance, retirement plans, etc. Most importantly, I want to have passion for my work, I want my work to help people and I hope to make a decent living while making the generations before me proud.

Another young woman made a similar point:

> I selected that I wanted autonomy out of working. I want flexibility—I want to work hard, but I also want to do many different projects and make various amounts of money doing so, while sometimes taking breaks from working at all that are more than just the 2 weeks provided in a year.

These data and comments suggest that members of today and tomorrow's work force share many of the same concerns of previous generations, perhaps giving more weight to making an impact and having enough flexibility to integrate their work and personal lives. MBAs who took the on-campus version of the course so they would be prepared to engage with the next-generation work force interpreted these data as suggesting that as managers they will need to listen to their employees, work together with them to solve problems that matter to the organization and to society, and to be flexible about how, when, and where people work so they can both be productive and attend to personal and family affairs.

What Can Be Done? Lessons from Three Decades of Research and Experience

Our ideas about what we think needs to be done, outlined below, come from over 30 years of research and direct involvement in the worlds of work and employment relations. Over this time much of our research, teaching, and work with government agencies, companies, and unions has had a common theme: the search for innovations that will improve both the performance of our economy and the quality of work and family life. We will draw on lessons learned from these experiences in the chapters that follow by presenting these ideas not as final solutions but as thought provokers and conversation starters. The goal is to encourage you to engage in this discussion and in doing so widen the circle of voices calling for change and working actively to make good things happen.

Some might argue that globalization and technological change or some other force of nature inevitably means that recent trends will continue. We would agree that this will be the case if we let the status quo continue unchallenged. But this is not inevitable—not if we act now in well-informed, coordinated, and strategic ways.

Yes, globalization and advances in technology are part of the forces at work today. But they are not really anything new. Similar disruptions and a similar need for fundamental changes in policies, institutions, and practices emerged when technological and economic advances accelerated the transition from an agricultural to an industrial economy in the early years of the twentieth century. During that transition, too, it was a long time before changes were put in place that allowed the work force to adjust and benefit from the new economy. Today we are well into another transition—*from an industrial economy to an innovation- and knowledge-based economy*—that is playing out on a global stage with perhaps even more far-reaching effects than the previous transition.

As in the early twentieth century, institutional inertia and political stalemate have resulted in a mismatch between changes in the economy, in the work force, and in the world of work and the policies, governance arrangements, institutions, and practices that govern work and employment relationships. Our task—indeed, our opportunity—is to create a better match.

The Foundation: Lifelong Education for All

What will it take to change this? For a start, nothing short of a fundamental set of changes in how we get educated, when we get educated, and who gets educated. The mantra has to be not quite (but almost) cradle-to-grave education for all. The rhetoric about lifelong learning will need to become a reality if the work force of the future is to compete on a global stage. And it has to start very early in life. There is clear and convincing evidence that investments in pre-kindergarten education pay off in the form of less crime, higher graduation rates, and greater earning potential. It is encouraging to see a consensus building around the need to broaden access to pre-kindergarten education in such diverse places as New York, Oklahoma, and Georgia.[3]

Then we have to keep up the pressure for innovation and change in elementary and secondary schools. The "Race to the Top" incentives and requirements the Obama administration created have spurred more innovation in public schools and more collaboration than ever between teachers, their unions, and school administrators who are focused on improving student achievement and school performance. Efforts are currently under way to expand these collaborative efforts in Massachusetts, New Jersey, and Illinois and the hope is that these initiatives will serve as models for the rest of the country.[4]

In his 2014 and 2015 State of the Union addresses, President Obama called for increased financial support and better coordination and improved results in the technical training provided to high school graduates and current members of the work force through community colleges and vocational schools and allied institutions. He went on to say "and we know how to do this." He's right. We've learned a great deal about how to make these work force development programs work—by getting employers in a community or industry to band together, work with these schools,

[3] For a summary of state-level pre-K initiatives, see Deborah J. Ackerman and Richard J. Coley, "State Pre-K Assessment Policies: Issues and Status," Policy Information Report, Educational Testing Service, 2012, accessed February 23, 2017, https://www.ets.org/Media/Research/pdf/PIC-PRE-K.pdf.

[4] For information on the New Jersey partnership effort, see "Labor-Management Conference Collaboration Showcases the Value of Educators' Voices," *Rutgers School of Management and Labor Relations*, accessed February 23, 2017, http://smlr.rutgers.edu/news/labor-management-collaboration-conference-showcases-value-educators-voices. For Massachusetts, see "Massachusetts Education Partnership," *Rennie Center*, accessed February 23, 2017 http://www.renniecenter.org/initiatives/labor-management-collaboration. For Illinois, see "About CEC," *Consortium for Educational Change*, accessed February 23, 2017, http://www.cecweb.org/.

provide on-the-job and classroom learning opportunities, and then provide jobs and career advancement opportunities to graduates. Making the funding for these programs contingent on showing that these features are in place would go a long way toward spreading high-quality programs across the country and eliminating any skill gaps that the business community says it now faces.

Then we come to that highly expensive yet valuable asset: a college degree. American universities are one of our national treasures—institutions that are still the best in the world. The problems are that they are beyond reach for too many and that debt burdens limit the ability of too many graduates to enjoy the benefits a college degree should provide. But fortuitous opportunities are unfolding in the form of universities around the world that are experimenting with various forms of online or distance-learning media. The challenge lies in translating the initial burst of innovation in online university courses into sustainable models that in fact provide access to lower-cost and higher-quality college experiences for students around the world. The online course that we are drawing on here is just one small experiment along these lines. Let's get many others in process and watch and learn from them as well.

These are the investments in education that business, government, labor, and educational institutions will need to make on a large scale to build the human and social capital American employers and workers need in order to be competitive in global markets. Individuals will also need to make on a lifelong—or at least a career-long—investment in ongoing education. Without a world-class education and world-class capabilities, individual workers will find it hard to compete with lower-cost workers in other countries and employers will be reluctant to invest in American jobs.

Using Technology to Complement and Augment Work

At the dawn of the industrial revolution in Britain, major technological advances threatened the livelihood of those who worked in the textile industry. This led some workers, who became known as Luddites, to attempt to destroy the new machines. Many technological scares of this type have come and gone over the years, always displacing some workers. These changes disrupt some individuals and their families, but they eventually spur further economic growth and create new job opportunities for workers with the right skills. We are facing this kind of situation in the wake of advances in digital technologies.

We will need to embrace advances in technology while we work to assure that they function to complement and augment, not eviscerate, work. We will need to adopt the mantra that it is "people who give wisdom to these machines," to borrow a Japanese phrase. Some experts predict, with good reason, that the coming wave of digital technologies will eliminate large numbers of good jobs. They are right, just as generations before ours worried about similar advances in technologies that freed people from laboring in farming and fishing; automated large numbers of difficult, repetitive, and sometimes dangerous production jobs; and transformed or eliminated office and clerical jobs such as chart-room workers in health care. Our challenge is to win the race with—not against—new technologies by inventing new ways to use them and by training workers and modifying work systems in ways that enhance the likelihood that technological investments will realize their full potential. In the best health care organizations, such as, for example, Kaiser Permanente, as technology changed how work was done, chart-room clerical workers were trained and transferred to new jobs that used electronic medical records technologies to promote preventive care and remind patients to get regular checkups.

We need to bring this type of new mindset to would-be inventors and designers of next-generation technologies. For too long, engineers have been trained to focus primarily on how to cut labor costs. While this has its benefits, so too would incentives to invent and develop new technologies—indeed, new entrepreneurial organizations—that use the skills of creative, dedicated people to tackle the globe's greatest environmental and human problems. Innovation and entrepreneurship are critical to job creation. The key is to make this a priority in our university, government, and private-sector laboratories and educational programs. If we combine this initiative with lifelong education programs and creative ideas for structuring work (discussed below), we will win the race with the new machines and put them to good use for humankind.

Good Businesses and Good Jobs

The values and business strategies that have dominated in most U.S. corporations in recent decades will need to change in big ways if the next-generation work force is to thrive and the environment is to be saved. For the past several decades, American business has narrowed its focus to a singular objective: maximizing short-term shareholder value. Some have labeled this the "financialization" of the American economy, arguing that pleasing Wall Street has essentially become the top priority of

U.S. firms. Business must be challenged to go beyond a focus on maximizing short-term shareholder interests to refocus on a more balanced set of objectives that involve people and the planet as well as profits. This will require overcoming the prevailing teaching about the role of the corporation that has dominated economics and business school curricula for the past two decades. An entire generation of managers and executives has been indoctrinated with the view that their primary, if not their sole, responsibility is to attend to shareholder interests and, even worse, to attend to and shape their own compensation and rewards to be aligned with short-term shareholder value.

It will also require a concerted effort to rebalance power in organizational decision making. Corporate norms don't rise or fall simply because of good intentions. Power also matters. How this shift in power will occur will be a major question for discussion and a major challenge for those who have the biggest stake in the future.

The good news is that we have learned how to do this directly from the businesses that have bucked this trend and focused on managing their affairs in ways that produce good financial results and good jobs and career opportunities. Every industry has its favorite examples. In the airline industry, it might be Southwest; in retail, it might be Costco; in software development, it might be SAS; in health care, it might be Kaiser Permanente. As noted earlier, the CEO of Market Basket, Arthur T. Demoulas, has now become the new poster child for advocating what have been called "high-road" or "high-performance" strategies and employment practices that work for multiple stakeholders. You may work for an organization like this. We need to learn what makes these examples successful for all stakeholders so that we then can get on with the task of turning them into the norm rather than the exception.

Worker Power—But Not Just Your Parents' Unions

You might be surprised, but in what follows we will argue strongly that the next generation will need its own sources of bargaining power to regain control of its destiny. In the post–World War II era, it was labor unions and collective bargaining that gave large number of workers the power they needed to improve working conditions, advance their standard of living, and build a strong middle class. Union advocates like to remind us (correctly) that they brought us the weekend and many other benefits we take for granted at work today. Most young people today have no knowledge of and perhaps little reason to even think about unions as anything other than relics of a bygone era. Yet as we will see, young workers will need to take collective actions if they are to turn the country around to meet their aspirations and expectations for good jobs, careers, and family lives.

But it can't be just your parent's or grandparent's labor movement—it will need to be a more nimble, more flexible, and less constrained set of networks and organizations that can empower workers to find good jobs, convince employers to meet their expectations, and provide opportunities for employees to update their capabilities throughout their working lives. Indeed, creating these new forms of advocacy, voice, representation, and support are some of the greatest opportunities and needs facing the next-generation work force. We will explore the experience to date with worker-centered apps that rate employers and jobs on the qualities young workers value most and identify which employers to avoid because they come up short on these qualities. This could be one new source of bargaining power for the new tech-savvy and mobile generation. Others will need to be invented as well.

This is not to say that all the strategies labor has used need to be abandoned. History suggests that doing so would only lead us to reinvent them at some point later on. And in fact, existing unions and professional associations and a variety of other worker advocacy groups are already developing and incubating new approaches and combining them with tried and true methods from the past. The key is to inform the current generation about what worked before and what lessons past experiences hold for the future and then let a thousand flowers bloom.

Government as a Catalyst for and Complement to Innovation and Good Jobs

One of the key lessons from our history is that, to paraphrase what House Speaker Tip O'Neill said in the 1980s, *all innovations are local.* (His exact phrase was "all politics are local.")

Almost all federal policies, including the cornerstones of the labor legislation passed as part of President Roosevelt's New Deal in the 1930s, were based on innovations that first were developed, tested, and carefully studied in state and local governments and/or the private sector. The good news is that a tremendous storehouse of innovation has emerged and been tested at these local and state levels in recent years. The task now is to wake up, indeed shake up, national policy makers who are in a position to put in place laws and guidelines that are well matched to the needs of the next-generation work force and economy.

The key is to make government a catalyst and a complement, not an obstacle, to innovation, change, and fairness in work and employment relationships. This is not rocket science, but it will require a groundswell of voices from the next-generation work force and their allies in business, labor, and education who want to step up to these challenges and opportunities. Nothing short of a wholesale set of changes in legislation and enforcement strategies, active cooperation from government policy makers, and a commitment of those on the front lines of business, labor, and education to enacting these changes in good faith will do the trick.

As one of the examples used at the beginning of this chapter suggested, a top priority will be to end the embarrassing situation of being the last highly developed economy and democracy to provide parents with the supports they need to meet their dual work and family responsibilities. From there we can work on updating wages and hours laws, labor relations laws, and other outmoded policies. And we can insist that government practice what it preaches by requiring government contractors who supply goods and services made in the United States and abroad to pay fair and acceptable wages and comply with accepted employment standards. An increase in the nation's minimum wage is also long overdue, as is demonstrated consistently in polling data and in the actions several states and cities have taken to increase their minimums, even though Washington remains deadlocked on this and other employment issues.

How to Make This Happen

We need to build on what we have learned about what works and what is broken in American workplaces, but information and even new ideas are only starting points for producing change. We need to alter the political discourse and build a broad coalition of voices and interest groups calling for changes that work for their specific needs and interests and for the common good. And we need to keep on innovating—bringing new ideas forward that are not chained to past practices, organizational arrangements, or traditions. Call this a "crowdsourcing" approach, if you will, or whatever new ways work for generating ideas from our collective wisdom.

This requires convincing these different voices and interest-group leaders to engage—perhaps to reengage—with each other in honest dialogue. Perhaps out of such discussions a common narrative will emerge—a vision for the future that otherwise divergent groups might rally around and use to drive change. This used to happen when labor was stronger and business leaders had a broader view of their roles and responsibilities. Together they forged and sustained the postwar social contract.

But over time, sad to say, cross-group dialogue largely stopped as business, labor, youth, and other groups all retreated into comfortable conversations with themselves.

We created an exercise to support this type of dialogue and tried it out in the online course. The students who participated in the exercise indicated what they expect business, labor, government, and education leaders to do to build and support a "Next Generation Social Contract." We will report their thoughts in the final chapter and put the exercise online for anyone to complete either individually or perhaps in discussions with leaders from these key groups. Perhaps these conversations will spur actions to put these ideas to work.

Hopeful Signs

Is this just pie-in-the-sky thinking or is progress possible? We are optimistic, even given the divisions the 2016 election uncovered. American society has responded to such crises before; it has a well-earned reputation for being a collection of pragmatic innovators rather than a group of ideological antagonists.

Another positive sign is the uptick in interest in entrepreneurship among students we teach and hear from around the country. Many in the current wave of hopeful entrepreneurs aspire to build mission-driven organizations that address significant social and/or environmental challenges in addition to experiencing financial success. This is extremely important and exciting, given that we know that new firms are an important source of innovation, new jobs, and business norms. Supporting this new wave of entrepreneurship will be an important part of how the next generation invents a new social contract that is appropriately matched to current realities.

A third source of optimism comes from the voices of the work force itself, both from the cross-section of young and mid-career workers who have spoken to us in our online classes to date and the voices we hear in talking with low-wage, immigrant, and blue-collar workers who worry they are being left behind, forgotten, or treated as scapegoats for our societal ills. We need to listen to these voices, and indeed we encourage all of you to engage with those who are doing the all-too-often invisible work that makes our communities and organizations thrive. Their voices need to be fully represented in any effort to build a new social contract.

As we take on these challenges it would serve us well to learn a bit from our history. In particular, we should look carefully at the last time our country faced equivalent challenges to both our economy and our democracy. We will do so in the next chapter by looking at how the nation survived the Great Depression of the 1930s and laid the foundation for what we will describe as the post–World War II social contract. We do so not to try to go back to those same policies and strategies but to heed philosopher George Santayana's warning that those who don't remember the past are doomed to repeat it.

Video Resources

The following videos covering topics in this chapter can be found at http://iwer.mit.edu/speak-up-for-work/:

- 3,000 Year History of Work in 3 Minutes
- A Message from Former U.S. Secretary of Labor Tom Perez
- A Message to HR Professionals
- The Millennial Generation's Views on Their World of Work
- Current Challenges and Opportunities: Today's Labor Market and Workplace in Perspective
- All Innovations Are Local

2

What Was the Postwar Social Contract, Where Did It Come From, and What Made It Work for Three Decades?

Throughout this book we use the concept of a social contract that was first developed by Jean-Jacques Rousseau and other philosophers. We do so in order to capture what we believe constitutes a social contract at work. By the social contract at work we mean the mutual expectations and obligations that workers, employers, and their communities and societies have regarding work and employment relationships. In this chapter, we describe the central features of the social contract that emerged from the New Deal labor legislation of the 1930s and took hold after the end of World War II.

Figure 1.3 captures the essence of that postwar social contract: Wages and productivity moved upward together from 1945 to about 1980 and in doing so helped expand the American middle class and ushered in a sustained era of broadly shared prosperity. Let's look at how this happened and see what lessons we might learn about how to build a new social contract tailored to our current and future economy and world of work.

In the 1920s, the economy was flourishing and business was booming, but the majority of Americans did not share in the wealth. Then, beginning in 1929, the economy entered the crisis of the Great Depression. The Roosevelt administration, paying heed to British economist John Maynard Keynes's macroeconomic theory, began spending public money to help get the economy back on track. This succeeded in stabilizing the economy. In addition, those most in need were helped in a major way by the enactment of a comprehensive set of policy reforms related to the labor market and labor relations.

These reforms became known as New Deal labor legislation. They created unemployment insurance, social security retirement and disability pensions, a federal minimum wage, and hours regulations that required employers to pay overtime premiums for work done beyond 40 hours per week. Finally, but perhaps most important for the longer run, it created a labor relations law and policy that enabled workers to join and sustain unions that were capable of bargaining for wages, hours, and working conditions and created a set of policies for resolving labor-management disputes (see Figure 2.1 for a summary). These achievements laid the foundation for a new social contract for the American economy and work force. But they were not enough to do the whole job. It took a set of actions on by workers, employers, unions, and government policy makers to build on this foundation during and after World War II to finally usher in what became known as the postwar social contract. This contract generally worked quite well for three decades (although less so for women and minorities). In hindsight, this period looks like a golden era for the American economy and the work force.

Figure 2.1. The New Deal Foundations

Four Pillars of the New Deal Labor Policy	What They Did
Unemployment Insurance	Provided income to unemployed workers for a temporary time with the expectation they would either be rehired or find a new job as economic conditions improved
Social Security and Disability Insurance	Provided retirement benefits to employees who had worked a minimum number of years and benefits to workers who became disabled and unable to work
National Labor Relations Act	Protected the right of workers to form independent unions and engage in collective bargaining
Fair Labor Standards Act	Established a national minimum wage and overtime pay requirements for a work week of more than 40 hours

The New Deal

Imagine it is 1930 and you are about to finish school and enter the labor force. What is going through your mind?

You came of age in the roaring 1920s, when the economy was booming. President Calvin Coolidge has said that the "business of the country is business," and the booming stock market seems to have proved his point. But somehow you don't feel so optimistic. Your family shared only a little bit in the economic growth of the mid-1920s (wages went up about 8 to 10 percent, but a giant portion of the gains went to the top 10 percent of the population). And whatever income gains your family made were quickly wiped out by the events following Black Tuesday in October 1929, the day the stock market crashed. By the end of 1930, real wages for average workers were no higher than they had been a decade earlier. Unemployment was 10 percent and rising rapidly. If your family was part of the 13 percent of the population that lived and worked on a farm, you were in even worse shape: You had steadily lost income throughout the 1920s even in the face of the business boom.

And then came the Great Depression. At its worst point, 25 percent of the work force was unemployed. Homelessness grew to the point that the name "Hoovervilles" was given to the communities of shanties homeless people created with whatever materials they could find. The point was clear: President Hoover was doing too little to combat the Depression. Farming families in Texas and Oklahoma who could no longer cope with the combined effects of the Depression and years of drought began the trek westward toward the promise of a better life in California. John Steinbeck vividly described this migration in *The Grapes of Wrath*. It all looked quite hopeless, and to those who valued our American way of life and political system, it looked quite dangerous. Radical insurrection seemed just around the corner!

Losing a job was disastrous in the 1920s and 1930s. There was no unemployment insurance and no health insurance. If you kept your job, your wages and working hours were likely cut. One report indicated that by 1933, nine out of ten companies had cut wages, 60 percent of the work force was working part time, and family income had dropped by 45 percent.

The accumulating combination of frustration and desire for change led to a political upheaval. The Republicans had controlled the White House for ten years, but in 1932, Democrat Franklin D. Roosevelt was elected and ushered in the New Deal.

While Roosevelt didn't have a clear agenda for change when he was elected, at least one of his advisors did. Frances Perkins famously warned the new president that if he chose her to be his secretary of labor, she would press for legislation to provide unemployment insurance, a program of retirement insurance and disability insurance, and a national minimum wage (see Box 2.1).

Where did Ms. Perkins get these revolutionary ideas? The answer to that question brings us to the first lesson that can inform efforts to build a new social contract.

There had already been a good deal of experimentation with the programs Frances Perkins outlined for President Roosevelt in progressive states such as Wisconsin, New York (where Perkins had been commissioner of the state department of labor when Roosevelt was governor), and Massachusetts. In fact, many of these programs were first developed by academics from the University of Wisconsin under the tutelage of Professor John R. Commons, who earned the title of "Father of the New Deal," since many of his students carried the ideas to Washington once Roosevelt and Perkins were in office.

Box 2.1. Frances Perkins's Vision and Agenda

Roosevelt came right to the point. "I've been thinking things over and I've decided I want you to be Secretary of Labor."

Since the call from his secretary, I had been going over arguments to convince him that he should not appoint me. . . . I said that if I accepted the position of Secretary of Labor I should want to do a great deal. I outlined a program of labor legislation and economic improvement. None of it was radical. It had all been tried in certain states and foreign countries. But I thought that Roosevelt might consider it too ambitious to be undertaken when the United States was deep in depression and unemployment. In broad terms, I proposed immediate federal aid to the states for direct unemployment relief, an extensive program of public works, a study and an approach to the establishment by federal law of minimum wages, maximum hours, true unemployment and old-age insurance, abolition of child labor, and the creation of a federal employment service.

The program received Roosevelt's hearty endorsement, and he told me he wanted me to carry it out.

Source: Frances Perkins, *The Roosevelt I Knew* (New York: Viking Press, 1946).

Consider, for example, the genesis of unemployment insurance and social security. This is how historian Arthur Schlesinger Jr. told the story. Shortly after taking office, President Roosevelt gave Frances Perkins the green light to work on the agenda she had laid out. She went to work on the idea of creating an unemployment insurance system by drawing heavily on experts from Wisconsin, most notably Paul Raushenbush and Elizabeth Brandeis Raushenbush (the daughter of Supreme Court justice Louis Brandeis), University of Wisconsin professor Edwin Witte, and Arthur Altmeyer. They developed a plan that called for state-level administration of unemployment insurance funded through a payroll tax that was prorated based on the number of unemployed workers in a firm experienced. After considerable debate over the technical details of this approach, the Roosevelt team adopted it and the president endorsed it.

At the same time, another group tackled the question of how to create an old-age insurance system and some means of providing for the families of workers who had died or had become permanently disabled. The president had already expressed his views to his secretary of labor on this issue. According to Schlesinger, Roosevelt said to Frances Perkins: "I see no reason why every child, from the day he is born, should not be a member of the social security system. . . . From the cradle to the grave they ought be in a social insurance system." Roosevelt went on to describe his views on how this insurance system should be financed: "If I have anything to say about it, it will be contributed . . . both on the part of the employer and the employee, on a sound actuarial basis. It means no money out of the Treasury."

The rest, as they say, is history. In January 1935, Roosevelt's social security and unemployment insurance bill was submitted to Congress. It was hotly contested, often in terms that should sound quite familiar to those who have followed the debates over the Affordable Care Act (popularly known as Obamacare).

A leading business group, the National Industrial Conference Board, said: "Unemployment insurance cannot be placed on a sound financial basis. It will facilitate ultimate socialist control of life and industry." Alfred Sloan of General Motors said, "The dangers are manifest." James L. Donnelly of the Illinois Manufacturers' Association insisted that the new bill would undermine the American way of life by "destroying initiative, discouraging thrift, and stifling individual responsibility." Republicans in Congress such as Representative John Taber of New York channeled these views: "Never in the history of the world has any measure been brought in here so insidiously designed as to prevent business recovery, to enslave workers, and to present any possibility of the employers providing work for the people." Representative Daniel A. Reed concurred: "The lash of the dictator will be felt and twenty-five million free American citizens will for the first time submit themselves to a fingerprint test."[1]

But in the end, after lengthy deliberations and a number of amendments, Congress enacted the Social Security Act of 1935. It provided unemployment

[1] Arthur M. Schlesinger Jr., *The Age of Roosevelt: The Coming of the New Deal, 1933–1935*, vol. 2 (Boston: Houghton, Mifflin, 1958), 311, 308.

insurance, old-age insurance, and disability insurance programs. Little did these policy makers or their supporters and critics in Congress know that some fifty years later, Republicans and Democrats alike would describe Social Security as the "third rail" of politics that was never to be touched.[2]

As controversial as this social legislation was, the most difficult part of the New Deal had yet to be enacted—legislation to protect workers' rights to join a union and engage in collective bargaining over their wages, hours, and working conditions.

Neither President Roosevelt nor his cabinet members initiated this legislation, and most of them didn't even support it at first. Instead, its chief sponsor was Senator Robert Wagner of New York. The "Wagner Act" (formally the National Labor Relations Act), which Congress passed in 1935, shared two similarities with other parts of the New Deal: It built on local-level innovations (in this case in the private-sector clothing, coal, and railroads industries) and it was informed by the work of labor economists and historians who had studied and help guide collective bargaining programs in the era before the New Deal.

One final piece of New Deal labor legislation was passed. The Fair Labor Standards Act of 1938 instituted the nation's first minimum wage (25 cents per hour), required employers to pay overtime for a work week of more than 44 hours (later lowered to 40 hours), and abolished most child labor. President Roosevelt strongly supported this legislation, which Secretary Perkins and her staff at the Department of Labor had developed. Business strongly opposed it. Labor leaders were lukewarm in their support, fearing in part that government-mandated minimum wages would undermine unions and collective bargaining. Secretary Perkins's staff developed additional plans to strengthen a law already on the books that required government contractors and employers in government-financed construction projects to pay "prevailing wages."

We present this history, focusing particularly on the staff work done at the Department of Labor under a strong and well-informed secretary, to contrast it with the inaction on labor policies in recent Democratic (and Republican) administrations.

[2] This term is attributed to Speaker of the House Tip O'Neil in 1981. See William Safire, "Third Rail," *New York Times Magazine*, February 18, 2007, accessed February 23, 2017, http://www.nytimes.com/2007/02/18/magazine/18wwlnsafire.t.html.

The Macro Engine for Growth

Roosevelt did not begin his efforts to cope with the Great Depression with labor legislation. His first and most urgent task when he took office in 1932 was to stabilize the financial system. First, he implemented a bank holiday to stop the run on withdrawals. Then he embarked on a spending program to try to regenerate economic growth. By 1937, he had partially succeeded. But those in his administration who feared inflation more than unemployment won out and took actions to raise interest rates and limit the inflow of gold into the country (our currency was still tied to the value of gold) and the country fell back into recession. It took the buildup for World War II and the subsequent wartime production to finally bring unemployment down to pre-Depression levels.

While the government spent enough in the 1930s to keep the economy from sinking further into decline and to contain the social chaos that threatened to replace our democratic government with a socialist, communist, or right-wing totalitarian alternative, it took the massive expenditures of World War II to finally pull the economy out of recession and get back to something close to full employment. At the same time, the wartime labor shortages brought in massive female labor force participation. Rosie the Riveter helped produce wartime goods and keep young families afloat while young men—husbands and fathers—went to war.

Again there is a lesson for today.

It took massive government spending to recover the jobs that had been lost in the Great Depression. After the war, the restored purchasing power of consumers sustained a strong labor market for many years. Today, might rebuilding our crumbling infrastructure be a substitute for war?

While the war buildup brought the unemployment problem under control, a new challenge emerged: How could the government keep wartime production levels going without work stoppages and without letting inflation get out of control? The answer turned out to be a little-known and underappreciated institution composed of government, business, and labor leaders. The National War Labor Board (NLWB) was important for creating the principles and practices that would help usher in decades of shared prosperity after the war.

The Government as an Innovator: The National War Labor Board

A number of newly minted PhD economists schooled in the latest developments in economic theory were brought in to staff various wartime agencies and to put their theoretical knowledge to work on the practical processes of collective bargaining, wage determination, and labor-management relations. Box 2.2 presents a quote from one the best known of these young economists, Clark Kerr, who went on to become one of the nation's leading mediators and arbitrators and, eventually, the president of the University of California.

> **Box 2.2. Clark Kerr's Story**
>
> When I entered the field of industrial relations, I had a chance to practice the art of peaceful solutions. My first experience in the field [while studying for my PhD] was in the fall of 1933 during the bloody cotton pickers' strike in the great Central Valley of California. Then, later[,] . . . from 1940 to 1945 I became the leading arbitrator of industrial disputes in the Seattle region. This led to my participation during World War II in the work of the regional War Labor Board stabilizing wages and settling labor disputes, hundreds of them. After the war, I continued in arbitration and became a leading arbitrator on the West Coast. I saw how violence once unleashed came to lead an uncontrolled life of its own. I saw how patience and reason led to less costly processes and better solutions than did passion and violence.
>
> Source: Clark Kerr, *The Gold and the Blue: A Personal Memoir of the University of California, 1949–1967*, vol. 1, *Academic Triumphs* (Berkeley: University of California Press, 2001), 14.

Members of the NWLB such as Clark Kerr and Professor George Taylor from the University of Pennsylvania helped invent and spread many of the employment practices that made it possible for professional personnel management and collective bargaining to work effectively for decades to come. Rational internal job structures, wages tied to levels of skill and responsibility, formulas that adjusted wages for changes in the cost of living, comparisons of wages within industries and occupations, fringe benefits including health insurance and pensions, grievance procedures that included arbitration to resolve day-to-day disputes—all of these grew out of the decisions or recommendations of the NWLB. And based on their experiences in their early careers in working with management and labor to apply these new principles, a cadre of young professional labor relations "neutrals" was created. These professionals represented neither labor nor management but worked with both sides as mediators or arbitrators to resolve their disputes. They went on to apply and adapt these practices in industry for decades.

The lesson: Creative and knowledgeable "neutrals" can make a difference and invent solutions to practical problems. But they have to understand the nature of labor market (and other) forces and not be captured by the interests of one party or another—in this case business or labor. And they can't be so tied to past practices that they are not able to invent new solutions for the future. Perhaps a new generation of neutrals is needed today to bring society's diverse stakeholders out of their isolated silos to work together to solve problems.

By far the most important of the NWLB innovations involved fringe benefits. During the war, the board encouraged management and labor unions to adopt or increase health insurance and pension plans as a way of holding wages in check and keeping labor peace. This is how employers came to be the providers of long-term economic security and health care coverage for their employees. The tactic worked well for many years, at least for those who were covered, although the legacy has over time become an albatross around the neck of the economy as more workers move in and out of coverage when they change jobs or end up without coverage in part-time jobs, work contract jobs, or self-employment.

The lesson: It made good sense to use individual employers as the transmission belt for spreading health insurance and pension coverage when large firms and long-term employment with a single firm was the dominant model of employment relations. This is no longer the case, and we now have to wean ourselves from this outmoded approach to funding and transmitting coverage of these key benefits. Portable benefits have to be part of the future of work.

The Postwar Economy and Labor Market: Boom or Bust?

After World War II, many economists worried that the economy would fall back to prewar levels of stagnation as government wartime expenditures declined. In 1943, economist Paul Samuelson, who later was awarded the Nobel Prize in economics, wrote that when the war ended, "some ten million men will be thrown on the labor market" leading to the potential for "the greatest period of unemployment and industrial dislocation which any economy has ever faced."[3] Another future Nobel Prize winner, one of the most famous European economists of the time, Gunner Myrdal, offered the even more dire warning that the economic challenges in Europe would lead to an "epidemic of violence."[4]

But lo and behold, neither Samuelson's nor Myrdal's prediction came to pass. Instead, during the postwar period, a robust economy emerged in the United States led by pent-up consumer demand and fueled by financing provided largely by the United States through what became known as the Marshall Plan (named after Secretary of State George Marshall) to support the rebuilding of the war-torn societies of Europe and Japan. This was achieved by a combination of the private-sector investment that was needed to help industry transition from producing military goods to producing consumer goods and a supportive set of education and labor market policies and institutions that matched the needs of the postwar economy.

Together, these business investments and institutions created what would become known as the postwar social contract. Here's how it was created and sustained for the three decades following the war.

The Postwar Social Contract

The Role of Education

Let's start with the role education played. When World War II ended, 10 million veterans who had put their careers on hold so they could serve their country returned home to relaunch their family lives and careers. Some, like Ted Williams, the most famous Red Sox player ever, were so talented that they could pick up where they left

[3] Paul Samuelson, "Full Employment after the War," in *Postwar Economic Problems*, edited by S. E. Harris (New York: McGraw-Hill, 1943), quoted in Cecil Bohanon, "Economic Recovery: Lessons from the Post World War II Period," September 10, 2012, http://mercatus.org/publication/economic-recovery-lessons-post-world-war-ii-period.

[4] Ibid, 279–280.

off. The first year he was back, Williams batted .342, hit 38 home runs, knocked in 123 runs, and led the Red Sox to the American League pennant (but, as all Red Sox fans know, not on to win the World Series; that would have to wait until 2004).

Other veterans were not yet at the top of their game and the nation worried about what to do with them. Fortunately, national leaders were also worried about the future of veterans and felt society owed them some assistance. For veterans who had grown up on family farms, going back was not a good option because advances in technology—tractors and milking machines, for example—were greatly reducing the need for farm labor and small family farms were becoming more and more tenuous and less likely to survive long enough to be passed on to the next generation. Fortunately, the manufacturing sector beckoned. Large industrial firms were growing and needed middle managers and talented technical engineers. But these opportunities required further education.

The GI Bill was created to meet these needs. By any standard (particularly compared to today), the benefits were generous. The Servicemen's Readjustment Act of 1944 (the official name of the GI Bill) entitled anyone with 90 days or more of military service to one year of tuition and paid fees for education up to a maximum of $500 per year. The time of support increased for each month of service, up to a maximum of four years. In addition, single veterans who were in school received a stipend of $50 a month; married veterans received $75 a month.

About 12 percent of returning veterans took advantage of the opportunity to go to college. One study estimates that the net effects of the GI Bill and military service in World War II was an increase in college graduation rates of 5 to 8 percent. Other studies estimate smaller effects. There is no question that the benefits were generous enough to provide a strong incentive for veterans to continue their education and that they provided enough support for veterans' families while they were studying that veterans could graduate with little or no debts to repay. These extensive benefits not only encouraged college attendance but very likely helped many veterans study at prestigious institutions that they otherwise would not have been able to attend. As just one example, over 90 percent of those admitted to the Harvard Business School in 1947 were supported by the GI Bill.

During the postwar years, American universities grew in size and stature to become part of the world's best and most accessible system of higher education. Our favorite example of a great public university is the University of Wisconsin. The relatively low cost of education at institutions such as this one enabled many young people to become the first one in their family to attend and graduate from college, often through a combination of the GI Bill, part-time work, scholarships, low-interest loans, and family support. One of us transferred to UW-Madison right out of the military, when tuition was about $300 per year, while the other one started college at the UW Manitowoc County Center (a two-year institution that provided transfer credits to the UW-Madison campus), where tuition was $210 per year. (In 2016, in-state tuition at the UW-Madison campus was $10,488 a year.)

The quality of this education was unsurpassed. While funding for higher education was always controversial in state politics, successive waves of state legislators and governors supported Wisconsin's state university with generous budgets. But this support has decreased markedly since 2005 and reached a nadir (we hope) with the budget cuts Governor Scott Walker imposed after he was elected in 2010. When we graduated from the university with our PhD degrees in 1972 and 1973, the state government covered 43 percent of the university's total budget; by 2015, the state's contribution had fallen to 17 percent. Over this same time period, the proportion of the university's income from tuition increased by nearly 50 percent, rising from 11 percent in 1973 to 16 percent.

We summarize the University of Wisconsin experience to illustrate the risks America is facing as it defunds public universities and makes it harder and harder for young, ambitious, and talented people from families of modest means to use these premier institutions as a channel for upward mobility. Although higher education remains one of the things America excels at, if the nation is to retain this position, significant transformation will be necessary in the years ahead. For this reason, we need to understand the role of education in fostering and supporting the postwar social contract, if only so we can figure out what features need to be retained and reinforced and what features need to be to changed and transformed in the future.

The lesson: Affordable college tuition is important for all young people but especially for families with first-generation children aspiring to get a college degree.

The Role of Collective Bargaining

In the postwar years, demand for production workers was brisk as factories retooled to meet the pent-up demand for consumer goods that had not been available during the war. Unions entered a period of growth and became permanent institutions in the United States, in part because the 1935 National Labor Relations Act required employers to bargain with unions. But workers also had pent-up demands following years of wage controls. As these controls were lifted, numerous strikes broke out. More time was lost to strikes in 1946 than any other year before or after. If any business leaders thought the end of the war would usher in a return to the pre-union conditions of the 1920s, the strike wave and the newfound power of industrial unions demonstrated that this was not an option. Instead, employers needed to come up with a way to stabilize labor relations so they could take advantage of growing markets for American goods at home and abroad. The postwar social contract emerged from this setting.

President Truman hoped that the cooperation between labor and business that had developed during World War II could be carried over to the peacetime economy. In 1945, he called a meeting of national labor and business leaders to discuss the principles of a potential postwar labor-management accord. Although the parties came close to such an agreement, they came up short on one issue: They could not agree on the limits of union influence in management decisions.

Progressive union leaders such as Walter Reuther, president of the United Auto Workers (UAW), envisioned a postwar labor relations system in which workers would contribute to improving production processes and help steer businesses toward broader social purposes. In the early 1940s, Reuther had led the process of retooling industry to build tanks and ships to support the war effort, and he had hopes that labor leaders could play a significant role again as the nation entered a new period of transition. But business leaders strongly opposed having an open-ended agenda for labor-management relations. They wanted to retain management's right to manage businesses.

The lesson: It is likely that there will never be a permanent accord between labor and business. But when the nation is in crisis, both sides can be mobilized to act in the national interest. At least this was the case in the past. What steps might be taken to rally the same sense of national solidarity to help forge a new social contract suited to today's economy and the work force of the future?

In the absence of a shared vision for the future of labor-management relations in the postwar years, the parties each had to pursue their own strategies. Walter Reuther, who, depending on one's point of view, was either the "most dangerous man in Detroit" or the most progressive labor leader of his time, pursued his vision through bargaining with auto companies.[5] Reuther wanted workers to have a voice in management decisions about production, pricing, and product development in addition to the legally prescribed range of bargaining over wages, hours, and working conditions. But management had a different idea, and few of Reuther's labor leader colleagues supported his ideas. Instead, GM's CEO proposed what the editors of *Fortune Magazine* labeled the Treaty of Detroit. In return for labor peace, GM would agree to a wage formula that would link wage increases to growth in productivity and to increases in the cost of living.[6]

This wage formula and other aspects of the settlements between GM and the UAW spread across the economy. It is worth reflecting on how this happened, not just for the historical record but because, as we will discuss later, a big strategic puzzle today is how to make the best practices of leading firms the norm in business. To understand how this happened, we need to introduce a labor relations term: pattern bargaining.

This is the process that unions use to spread wage settlements achieved in one firm to competing firms in their industry or labor market. The idea is to take wages out of competition; that is, to work toward similar wage levels for similar work across an industry. By doing so, the unions provide stability in labor relations and gradually raise or ratchet up the floor on wages, benefits, and working conditions. In today's parlance, we might describe this as a way of avoiding a race to the bottom.

For example, the UAW took the Treaty of Detroit agreement to GM's direct competitors, Ford and Chrysler, and was able to get them to agree to the same settlement. Unions in other industries such as steel, electronics, oil and gas, utilities, and rubber likewise sought to match what the auto workers achieved. The net result was the tandem movement in wages and productivity from the mid-1940s through the 1970s that is depicted in Figure 1.3.

[5] George Romney, the future governor of Michigan and the father of Mitt Romney, the Republican candidate for U.S. president in 2012, is the one who called Reuther "the most dangerous man in Detroit." See Nelson Lichtenstein, *The Most Dangerous Man in Detroit: Walter Reuther and the Fate of American Labor* (New York: Basic Books, 1995), 230.

[6] Ibid., 279–280.

Econometric evidence has demonstrated that collective bargaining was a particularly effective way to use pattern bargaining to reduce wage differences within industries. Although there was no single national wage pattern, a norm developed by which unions bargained for wages and employers agreed to wage proposals based on comparisons with what was paid by similar-sized competitors in the same industry. This norm was enforced by the bargaining power of unions and was copied by nonunion firms that wanted to remain nonunion. That is how wage increases among blue-collar workers were spread across competitors within industries and labor markets. At the same time, the salaries of white-collar workers and middle managers also increased because personnel managers were careful to maintain reasonable differentials between managers' wages and the wages of the workers they supervised. The salaries of CEOs and top management were held in check for fear that unions would demand equivalent pay increases if they observed those at the top of their companies disproportionately feathering their own nests. The combination of union power and pattern bargaining created social norms that kept wage and income inequality from getting out of hand in the postwar period.

We polled the students in our online class at MIT about whether they thought wages and productivity should move together as they did in the heyday of the postwar social contract. An overwhelming majority—83 percent—agreed that they should. They saw it as a sensible general norm or principle for wage setting. But many also worried that this might be difficult to do in the future. Here is a sampling of student comments on this issue.

> The lack of congruence between wages and productivity [in recent decades] is one of the reasons we have such drastic income inequality. If the benefits of all that productivity are not going to the workers, then they are going to the owners. Thus the owners, i.e., the major stockholders, are becoming enriched at the expense of the workers who are creating the wealth for them.

> Shareholders and managers have always been and always will be motivated to increase profits, which means paying workers the lowest possible wages. In the postwar era, workers had the power (mostly through unions but government also played a role) to force employers to pay higher wages. Since the 1970s workers have lost that power. To increase wages, workers must regain the power to force wage increases. Wage increases will not happen through employer charity.

The lesson: Norms don't appear out of thin air. Behind every norm lies an idea and the power to enforce and spread that idea. What will be the new wage norm in the future and where will the power come from to enforce it?

From Steady State to Atrophy

All these cross-cutting institutions helped support and sustain the social contract from the 1950s through the 1970s. During these years, as the lines in Figure 1.3 indicate, the wage-setting formula that GM and the UAW initiated in the 1940s kept productivity and real wages moving upward roughly in tandem. This is not to say that there were not rough spots along the way. In the 1960s, concerns that wage-price spirals were fueling inflation led the Kennedy and Johnson administrations to introduce wage and price guideposts in an attempt to restrain inflation. In the early 1970s, runaway wage increases in the construction industry that threatened to spread to other industries led the Nixon administration to take even stronger action in the form of wage and price controls. And later in the 1970s a period of "stagflation"—slow economic growth while wages and prices continued to increase—created a crisis that eventually led to dramatic change in both economic policies and political leadership. The postwar social contract had matured and was failing to adapt to an incrementally changing environment.

Indeed, the 1960s proved to be a tumultuous decade in both employment relations and American society in general. In employment relations, the 1960s began with much concern that advances in technology (referred to at the time as automation) were gradually but steadily eliminating jobs and creating a population of permanently unemployed workers (called structural unemployment). A host of new policies were implemented to support retraining, geographic relocation, and regional economic development to cope with the consequences of persistent unemployment. The employment and training policies and infrastructure that are in place today are essentially carryovers from these early steps toward a national labor market policy.

The automation scare proved to be overstated and premature. Just as World War II expenditures brought the labor market out of the Great Depression, expenditures for the Vietnam War in the 1960s did more to bring down the unemployment rate than the new employment and training policies did, although they were also helped along by the technological innovations that spawned the growth of the emerging high-tech industries. Once again the lesson is clear:

When an economy needs to create new, high-quality jobs, it must have strong growth-oriented macroeconomic policies in place and must nurture technological invention, entrepreneurship, and innovations. Will the digital innovations of today and tomorrow create enough new good-quality jobs or do we need to complement what these technologies are generating with greater macroeconomic investments in infrastructure, manufacturing, and education?

The social upheaval the Vietnam War spawned and the civil rights battles of the 1960s began to create schisms in the fabric of the social contract. Coming of age and entering the labor force in the 1960s was a heady experience. Everyone was fighting with everyone. The civil rights movement took off with marches in Selma, protests in Birmingham, and the famous March on Washington, where Martin Luther King Jr. gave his "I Have a Dream" address. The Vietnam War tore the country apart, and student protests at leading universities brought police and the National Guard to campuses across the country, in some cases (e.g., Kent State University) with tragic consequences. Cities such as Los Angeles and Detroit were literally on fire as the result of civil rights riots. Young people became disillusioned with all major institutions—with labor unions for being "hardhats" who supported the war and resisted integration, with businesses for making napalm and other horrific war materials, with university leaders for being part of the establishment. American society seemed to be coming apart.

While college students opposed the war, the most visible leaders of the labor movement and the business community either continued to support it or kept their personal misgivings to themselves. While some labor and business leaders supported civil rights activism, the most visible leaders— particularly the leaders of the AFL-CIO—remained silent or aloof. George Meany, president of the AFL-CIO, chose to be out of town the day of Martin Luther King's March on Washington in 1963, leaving his rival, Walter Reuther, to be the highest-ranking labor leader to march with Rev. King. Young people saw unions as so entrenched in the "establishment" that they couldn't possibly have much to offer the next generation.

Meanwhile, the world of work was changing below the sight lines of both established labor and management. New ideas for organizing work in more flexible ways had begun to emerge that allowed individuals and teams to flourish and informed how work was done, especially in the new high-tech industries represented by companies such as Hewlett Packard, Texas Instruments, Digital Equipment Corporation, and, later, Intel, Apple, Dell, and their progeny. These companies used new ideas to organize work, motivate employees, and provide a satisfying and challenging work environment. Labor unions remained stuck in organizing models that assumed that workers would be dissatisfied with their jobs and distrustful of their bosses and failed to adapt in ways that convinced workers in these emerging industries that they needed union representation. Gradually, the firms and unions that occupied the high-road cell in Figure 1.1 in the postwar era were migrating in the direction of the uncompetitive cell: they paid high wages but their profits and competitiveness were declining. As a result, throughout the 1960s and 1970s, union membership began what would turn out to be a long-term decline.

By the mid-1970s, the divide between the unionized sector of the economy that carried forward the wage formulas and work practices of the earlier era and the newer, faster-growing high-tech sectors of the economy was apparent. The difference between union and nonunion wages had increased from about 5 to 10 percent in the 1950s and 1960s to an average of 20 percent by the mid-1970s—a differential that caused employers with unionized workforces to cut jobs and to become increasingly concerned about their ability to compete. The pressures for significant change were building.

The most visible political warning signal—really a shot across the bow of labor-management relations—came in 1977–1978, during the Carter administration, when a mild form of labor law reform (the Labor Law Reform Act of 1978) that was backed by the labor movement failed in Congress. The business community was emboldened by the experience of blocking this reform in a government led by a Democratic president and Congress.

Economic warning signs were equally ominous. The stagflation of the 1970s doomed the Carter administration. It took the shock of a movie-star president to change the course of history, a history we are still living today.

In the 1980s, the postwar social contract broke down in dramatic fashion. It has yet to be replaced by one that better matches the needs of the economy or the future of work. Our task, which we discuss in the chapters that follow, is to understand how the environment for work has changed and how these changes are likely to shape the future of work and then to challenge each of the key stakeholders to step up to their responsibilities and opportunities to do what is needed, individually and collectively, to lay a foundation and provide the leadership needed to get this job done.

Video Resources

The following videos covering topics in this chapter can be found at http://iwer.mit.edu/speak-up-for-work/:

What Do We Mean by the Term Social Contract?
The New Deal
The Postwar Social Contract

3

The Demise of the Old Social Contract and the Challenges and Opportunities That Lie Ahead

The postwar social contract was built on a foundation of a mutually supportive set of economic and political conditions. Large oligopolistic and heavily regulated companies had the power to influence buying habits and set prices high enough to ensure satisfactory profits. Strong labor unions were able to secure consistently higher pay and better benefits for their members and for other workers. During this period, many employees were satisfied, business and labor leaders could be counted on to consider the public good in their decisions and actions when necessary, and public policy makers were hands off—they had no desire to upset the applecart. By the late 1970s, however, this remarkably coherent system had begun to fray in ways have reverberated throughout the economy ever since. By the 1980s, in the face of rampant global competition, rapid deregulation, and demanding stockholders, the old social contract had disintegrated. Employers, faced with intense competition and little opposition, initiated a series of actions that collectively undid the equitable distribution of incomes that earlier had lifted so many comfortably into the middle class and, in the process, restored their faith in the American Dream.

What follows is part of the story of how this all unfolded and where it has left us. But this is by no means the whole story. The more pressing issue is where do we go from here? In that vein, then, we go on to consider the challenges and opportunities that lie ahead, since these serve as the backdrop for the collective actions that will be needed as we strive to forge a new and better social contract and once again revive the American Dream.

The Demise of the Postwar Social Contract

Beginning in the late 1970s, the postwar social contract came under relentless pressure from the confluence of four major economic developments: globalization, deregulation, financialization, and automation.

Globalization

In the postwar period, the economies of Europe and Japan were in the process of rebuilding while the economies of many other countries had turned inward. U.S. firms were dominant, not only at home but also abroad. Since that time, however, globalization—the worldwide movement of economic activity—has increased steadily as businesses in Europe and Asia recovered and began penetrating U.S. and global markets. In turn, U.S. firms extended their reach into these rejuvenated economies and, later, into newly developing countries. World trade—that is, the sum of exports and imports—grew rapidly, from 30 percent of gross domestic product in 1973 to 60 percent of gross domestic product at its peak in 2008. Initially, the growth was in so-called interindustry trade. Countries mostly exported products and services that were different from and thus not in direct competition with those they imported. More recently, however, the volume of intraindustry trade has increased, which had the effect of bringing more and more companies from more and more countries into one another's product and labor markets. Further, in recent years, the growth in world trade has been outpaced by foreign direct investments that include facilities firms have built in locations other than their home countries in order to take advantage of proximity to customers or low wages or both. Auto companies from around the world—such as Toyota, Nissan, Honda, Volkswagen, and BMW—have built over 300 facilities of various types in the United States. General Motors, Ford, and Chrysler (now Fiat Chrysler) have done the same elsewhere, albeit at lower levels, most notably in Mexico.

As these developments unfolded, large U.S. companies in particular found themselves in increasingly precarious competitive positions. In part, the pressure came from a wave of companies from both developed and developing economies. Having reached massive scales at home, these firms began aggressively pursuing additional opportunities elsewhere. And in part the pressure came from smaller, firms (often high-tech companies) that have pursued niche markets where larger, less agile firms were vulnerable. Consumers worldwide soon learned that it was possible to obtain, indeed to demand, state-of-the-art products and services of high quality at affordable prices. To gain (or at least hold) ground, large and many middle-sized U.S. firms rather quickly morphed into global enterprises, first by fragmenting their operations and distributing them to wherever the work could be done best or cheapest and then by taking advantage of emerging information and communication technologies (e.g., the Internet) to weave these facilities together into sophisticated global supply chains. Often the chains were broadened to include multiple companies in multiple countries. Walmart, for example, became the major player in retailing in large part because of an intricate supply chain that enabled the company to purchase goods from (and beat down the prices of) a host of smaller suppliers worldwide. In this increasingly dynamic and highly competitive world, several well-known firms in various industries failed. (Anyone remember Compaq or DEC computers, Woolworth's "five and dime" stores, Standard Oil gas stations, or Kodak Cameras?) Many others, however, managed to revamp their operations and do just fine for their owners and executives, if not always for their employees.

Deregulation

As globalization was unfolding, a wave of deregulation swept the United States. The financial sector was unbundled in a series of steps between 1974 and 1982: airlines in 1978, trucking and railroads in 1980, and telecommunications in 1982. This happened in two ways. In the financial sector, for example, big banks took the lead in an effort to gain access to customers in local communities, where smaller banks had protected markets. They were later joined by pension funds, mutual funds and insurance companies in this endeavor. In telecommunications, the pressure came from smaller firms with advanced technologies that wanted access to AT&T's long-distance network, which controlled telephone services in many local markets. In both cases, the result was the same. Firms that had for decades controlled consumers' choices and the prices charged for the goods and services they provided found themselves competing for business with smaller, more agile firms that had none of the legacy costs associated with outdated facilities and equipment, not to mention the generous wage and benefits packages that had accumulated over the years in the larger firms. Turmoil reigned in these deregulated industries. Start-ups came and often went in rapid succession. (Has anyone ever heard of People Express Airlines? It was the darling of the business press in 1980, but by 1987 it had gone bust.) Many of the previously regulated firms failed to adapt and disappeared. Others struggled but eventually regained their footing and became more competitive. The overall result was better products and services (although this was not always the case, as airline passengers can attest) and lower prices for consumers. But here, too, it was employees who often paid the price.

Financialization

Financialization is a broad term that is often used to describe the process by which the financial sector (i.e., Wall Street) gains size and influence over the strategies of individual firms and in some cases entire national economies. In the United States, it unfolded after the deregulation of the financial industry. The consequences have been numerous and costly.[1] Financialization's grip on business cannot be underestimated. In a recent survey of chief financial officers, for example, 78 percent indicated that they would forego economic value and 55 percent said they would cancel a project with a potentially positive financial payoff—that is, that they would willingly sacrifice their companies' long-term prospects—to meet Wall Street's demand for steady earnings.[2]

One particular aspect of financialization, the process by which the interests of firm owners and stockholders came to dominate the concerns of all other stakeholders, including most especially employees, is what we are interested in here. In the postwar period, most stockholders were wealthy individuals who were doing just fine and thus had little incentive to question how the companies they invested in did business. Over time, however, these individuals began turning their portfolios over to big investment banks and to hedge funds and the like to be actively managed, while at the same time smaller investors became more prominent as their cash began piling up in pension funds and mutual funds. Only 20 percent of the U.S. population owned stock in 1985; by 2005, that number had reached 50 percent. Consequently, financial institutions came under constant pressure to do well by their growing groups of clients and, in turn, leaned heavily on the companies in which they invested to deliver ever-increasing profits on a regular and short-term (quarterly) basis.

Wall Street bankers and their peers in the financial industry used a carrot-and-stick approach. First, to make sure executives were on board, they encouraged the development of pay plans in which large portions of executives' earnings were in the form of stock options and/or bonuses tied to stock prices. And in cases where this failed to elicit the desired results, they simply put pressure on companies' boards of directors to fire the executives involved. (However, this was often more of a threat than a reality. Performance-related sackings among CEOs are relative rare—only between 3 and 7 percent, depending on who's counting.)

The outcomes were predictable. The financial firms became larger and larger and thus more and more powerful, not to mention immensely profitable for their owners and lucrative for their executives. Corporate executives who consistently

1 Rana Foroohar, "American Capitalism's Great Crisis," *Time*, May 12, 2006, accessed February 23, 2017, http://time.com/4327419/american-capitalisms-great-crisis/.

2 Gautam Mukunda, "The Price of Wall Street's Power," *Harvard Business Review*, June 2014, accessed February 23, 2017, https://hbr.org/2014/06/the-price-of-wall-streets-power.

delivered ever-ascending profits garnered pay packages totaling multiple millions of dollars a year. All of this was occurring as globalization and deregulation were making profits increasingly difficult to come by. Managements took many steps to make their firms more innovative and agile in the marketplace and thus more responsive to customers' ever-changing needs. But of course it also was essential to become more efficient and to reduce costs wherever possible. For most firms, labor costs are significant, so they by no means escaped notice. Some firms, as we shall see in the next chapter, took the high road, searching for ways to both attain better financial returns for owners and stockholders and secure employment and high-quality jobs for employees. But in most cases the financiers and firms were unaware of or unappreciative of the high-road approach. They knew they had the power to discipline management and were not at all hesitant to use it.

The Four Major Responses

In response to these pressures, employers instituted four dominant responses during this time: 1) the containment of wages, salaries, and benefits; 2) domestic outsourcing; 3) a combination of global outsourcing and offshoring; and 4) the automation of work processes.

Containment of Wages, Salaries, and Benefits

This is the most obvious development. As noted previously, the average pay and the average total earnings of American workers have remained virtually flat for the past three and a half decades. Simply put, companies haven't paid more because they haven't had to. The power of employees to demand pay increases dissipated in the face of slack labor markets (between 1980 and 2010, the U.S. experienced four recessions), the declining prevalence and bargaining power of labor unions, and a failure to increase the federal minimum wage. Consequently, most of the gains generated by increasing rates of productivity went to consumers in the form of lower prices and to the owners of capital in the form of increased stock prices and dividends. Recall that picture of the divergent trends in productivity and wages shown in Figure 1.2 of Chapter. 1. And of course much of the increase in profits went to the executives who make pay decisions for their companies. In 1980, the ratio of CEO pay to the pay of the average worker in their companies was 40 to 1; today it is well over 300 to 1.[3]

Pay cuts were relatively rare, although not unheard of, but employers found several other ways to keep a lid on earnings. Obviously, these included modest (and sometimes no) budget allocations for pay increases. Employers also made efforts to convert pay from a fixed to a variable cost. Annual across-the-board pay increases gave way to more rigorous merit pay plans that grant generous increases to a few high performers but little to anyone else. It also became increasingly common for employers to grant merit awards in the form of one-time bonuses that don't accrue from year to year instead of in the form of actual pay increases that do compound over time. According to a survey by Aon Consulting, the number of performance-based bonus plans in the United States has jumped 18 percent since 2008. These bonuses currently constitute about 15 percent of total salary budgets. Less common, except in some unionized settings, was the adoption of two-tier wage systems in which some employees (often those who had just recently been hired) were paid less than the rate other employees earned for doing the same work.

In addition to pay, companies took drastic steps to control the costs of benefits, most especially pensions and health insurance. Not all firms, not even all large firms, provide pension plans for their employees. Many of those that do have switched from defined benefit plans to defined contribution plans (aka 401Ks) in recent years. A survey by Towers Watson, for example, found that in 1998 almost 60 percent of the Fortune 500 firms (292) had defined benefit plans for at least some of their employees. But by 2015 this number had dropped to under 20 percent (99) (mostly in the insurance industry and in public utilities).[4] The reason for the shift, of course, is that defined benefits plans require employers to provide predetermined payouts to retirees and thus involve long-term fixed obligations. Defined contribution plans, however, carry no such obligations since employees assume responsibility for making

[3] Lawrence Mishel and Alyssa Davis, "Top CEOs Make 300 Times More Than Typical Workers," *Economic Policy Institute*, June 21, 2015, accessed February 23, 2017, http://www.epi.org/publication/top-ceos-make-300-times-more-than-workers-pay-growth-surpasses-market-gains-and-the-rest-of-the-0-1-percent/.

[4] Brendan McFarland, "Defined Contribution Plan Volatility: Timing Is Everything," *Willis Towers Watson*, September 23, 2015, accessed February 23, 2017, https://www.towerswatson.com/en-US/Insights/Newsletters/Americas/insider/2015/09/defined-contribution-plan-volatility-timing-is-everything.

contributions to the plans and for making wise decisions about how their contributions to those funds are invested over the years. The fewer the contributions and/or the worse the decisions, the lower the payouts. Employers may match all of some of employees' contributions and may help employees find good investment advisors, but they aren't required to either. The same risky shift from employers to employees has occurred with respect to health insurance. Even before the passage of the Affordable Care Act in 2010, the number of firms providing health insurance for their employees was dropping steadily and plans in effect since that time have increasingly required employees to pay a greater share of the premiums and to absorb the costs associated with larger deductibles and co-pays.

Domestic Outsourcing

This is the process by which so-called host firms contract out work to other firms (or sometimes individuals) in the same country. Usually the contractors are smaller and more specialized companies. In some cases, domestic outsourcing is done as a means of gaining access to competencies (knowledge and skills) that are not available in the host company, while in other cases companies use it as a means of accomplishing work at a lower cost. Although domestic outsourcing has been around for a long time, it has grown rapidly in recent years, in large part due to the development of information and communication technologies that facilitate and lower the cost of coordination between host firms and their contractors and make it easier for host firms to monitor and maintain control over the quality of work that is done. Although good data are lacking, domestic outsourcing seems to be quite prevalent and appears to involve somewhere between 7 and 11 percent of the U.S. labor force.[5]

Domestic outsourcing aimed at competency acquisition is fairly benign in terms of the issues of concern here. In most cases, it is simply a matter of companies contracting through various types of service providers (e.g., research and development, consulting, or architectural firms) or online platforms such as Upwork

[5] Annette Bernhardt, Rosemary Batt, Susan Houseman, and Eileen Appelbaum, "Domestic Outsourcing in the U.S.: A Research Agenda to Assess Trends and Effects on Job Quality," IRLE Working Paper no. 102-16, February 2016, Institute for Research on Labor and Employment, University of California Berkeley, accessed February 23, 2017, http://irle.berkeley.edu/files/2016/Domestic-Outsourcing-in-the-US.pdf; Lawrence F. Katz and Alan B. Kreuger, "The Rise and Nature of Alternative Work Arrangements in the United States, 1995–2015," unpublished paper, March 29, 2016, accessed February 23, 2017, https://krueger.princeton.edu/sites/default/files/akrueger/files/katz_krueger_cws_-_march_29_20165.pdf.

for the services of highly trained professionals who tend to be relatively well paid. Domestic outsourcing in pursuit of lower costs, however, typically involves distributing what companies regard as noncore work (e.g., call center work, the administration of employee benefits plans, or the running of company cafeterias) to outfits that specialize in these types of activities. If the contractors are reputable firms that subscribe to high-road employment practices, then of course some employees lose jobs while others gain them, but overall the earnings effects are neutral or even beneficial for the employees involved. Often, however, this is not the case. Research has shown that in too many instances these kinds of outsourcing firms have been found to offer low pay—by sometimes as much as 15 to 20 percent—few benefits, and substandard working conditions.[6] Further, there are many instances where outsourcing is piled on top of outsourcing (in what have become known as fissured workplaces) to the point where the companies at the end of the chains feel pressure to engage in low-road work practices that sometimes include egregious violations of labor and employment laws such as the failure to pay overtime rates and even wage theft.[7] Thus, it is clear that domestic outsourcing in pursuit of lower costs has at least to some extent contributed to the wage stagnation and associated inequities that we documented in Chapter 1.

Global Outsourcing and Offshoring

These are the key elements of globalization. Both serve to redistribute work from one country to another. The difference between them is that global outsourcing involves two or more different firms (an example is Walmart's global supply chain), while offshoring refers to situations where the work is retained within the same firm (as is the case when a U.S. manufacturer builds, owns, and operates a plant in Mexico or China). Global outsourcing and offshoring can be undertaken to gain access to essential competencies not available domestically and/or in pursuit of low-wage workers. In some instances, this involves new work that could be done domestically but instead is sent overseas, in which case it means that the host country loses out in terms of job creation. Nike, for example, chose right from the start to have its shoes

[6] Bernhardt et al., "Domestic Outsourcing in the U.S.: A Research Agenda to Assess Trends and Effects on Job Quality."

[7] David Weil, *The Fissured Workplace: Why Work Became So Bad for So Many and What Can Be Done to Improve It* (Cambridge, MA: Harvard University Press, 2017).

and other products manufactured by contractors whose facilities are located in China, Indonesia, Thailand, Vietnam, and other countries.

But far more frequent and certainly more visible are the instances in which global outsourcing and offshoring involve exporting work that was once done at home to other countries. All kinds of work are involved here, including call center jobs, accounting, tax preparation, Web design, computer programming, data entry, legal services, routine healthcare tasks (e.g., reading X rays), and, most conspicuously, manufacturing.

The global outsourcing and offshoring of existing work typically results in layoffs. Here the research is quite clear; employees who are laid off from their jobs, irrespective of skill levels, generally experience at least some and often extended periods of unemployment. And more often than not when they are reemployed (really, *if* they are reemployed, since many become discouraged and drop out of the labor force entirely) they wind up taking jobs that offer lower pay, fewer benefits, and generally less desirable working conditions. A number of studies have shown, for example, that workers laid off from manufacturing jobs who subsequently become reemployed earn on average 20 percent less than they earned before they were laid off.[8] Thus, the global outsourcing and offshoring of existing jobs has the effect not only of adding to the rolls of unemployed and discouraged workers but also of suppressing the overall earning levels of those who are fortunate enough to find new jobs. These effects are vividly brought to life in media stories (and in politicians' speeches) depicting the total despair in city after city in America's "rust belt," an area of the country that has been decimated by the closing of numerous factories as work has been shipped overseas, particularly to low-wage countries.[9]

Automation

Since 1980, we have seen a rapid and impressive succession of developments in commercial applications of robotics; information and communication technologies, especially computers; the Internet; and software applications involving artificial

[8] Harry J. Holzer, Julia I. Lane, David B. Rosenblum, and Fredrik Andersson, *Where Are All the Good Jobs Going? What National and Local Job Quality and Dynamics Mean for U.S. Workers* (New York: Russell Sage Foundation, 2011).

[9] Derek Thompson, "A World without Work," *The Atlantic,* July/August 2015, 51–61, accessed February 23, 2017, https://www.theatlantic.com/magazine/archive/2015/07/world-without-work/395294/.

intelligence (AI) and machine learning. Over that time, corporate spending on automation outpaced investments in other capital equipment by a wide margin. Economists have long debated the extent to which automation affects patterns of employment and earnings.

Firms can choose to use automation in one of two ways. One—called *substitution*—is to use automation to replace entire jobs or most of the tasks associated with certain jobs. This is what will happen, for example, if self-driving trucks someday are used to displace the approximately three and a half million truck drivers now on the road. The other way firms use automation is referred to as *augmentation*. Here technology is used to supplement work in a way that makes it more efficient and/or better in some way. We can already see this process in work in the trucking industry, where many self-driving technologies are slowly being adopted as ways of make driving easier and much safer.

Since 1980, substitution and augmentation have had differential effects on different occupations. To clarify how this has worked, it is helpful to categorize work as, on the one hand, either routine or non-routine and, on the other hand, as either cognitive or manual.[10] Routine work consists of tasks that can be codified and reduced to fairly well-defined procedures, while nonroutine work consists of tasks that are more discretionary in nature. Cognitive work involves tasks such as problem-solving, reasoning, communication, high-level judgment, and persuasion. Manual work, in turn, consists of tasks that may involve interpersonal interactions but nearly always involve important physical components. Figure 3.1 provides specific examples of the tasks that fall in each of the four categories.

[10] D. Autor, F. Levy, and R. Murnane, "The Skill Content of Recent Technological Change: An Empirical Exploration," *Quarterly Journal of Economics*, 118, no. 4 (2003): 1279–1333.

Figure 3.1. Predictions of the Task Model for the Impact of Computerization on Four Categories of Workplace Tasks

	Routine Tasks	Nonroutine Tasks
	Analytical and interactive tasks	
Examples	Record-keeping Calculation Repetitive customer service (e.g., bank teller)	Forming/testing hypotheses Medical diagnosis Legal writing Persuading/selling Managing others
Computer impact	***Substantial substitution***	***Strong complementarities***
	Manual Tasks	
Examples	Picking or sorting Repetitive assembly	Janitorial services Truck driving
Computer impact	***Substantial substitution***	***Limited opportunities for substitution or complementarity***

Source: D. Autor, F. Levy, and R. Murnane, "The Skill Content of Recent Technological Change: An Empirical Exploration," *Quarterly Journal of Economics*, 118, no. 4 (2003): 1286.

Extensive analysis done by Professor David Autor of MIT and his colleagues shows quite clearly that during the past thirty-five years, automation has had a polarizing effect on occupations.[11] Specifically, it has significantly reduced the number of people who primarily perform routine tasks, both cognitive and manual. The biggest effects have been on the number of workers in clerical jobs (down 8 percent), transportation and construction jobs (down 15 percent), and, especially, manufacturing jobs (down between 38 and 54 percent, depending on the skills involved). This isn't unexpected, since many of the tasks that make up these jobs are precisely the kinds of things computers and robots are very good at. To date, automation has largely served to augment the work of those who do primarily nonroutine, cognitive tasks, including technicians, professionals, and managers, and

[11] John Danaher, "Automation and Income Inequality: Understanding the Polarisation Effect. Institute for Ethics and Emerging Technologies," *Institute for Ethics and Emerging Technologies*, October 10, 2015, accessed February 23, 2017, http://ieet.org/index.php/IEET/more/danaher20151010.

these occupations have all increased their share of employment over time (by 28 percent). Finally, automation thus far has barely affected those who perform primarily nonroutine, manual tasks. These workers—waitresses and others in food services, home health aides, security guards, and the like—do tasks that involve things such as situational awareness, interpersonal interactions, and physical adeptness that computers have yet to master. Employment in these occupations also has grown quite rapidly (up 30 percent—about the same as occupations that primarily involving nonroutine, cognitive tasks).

But has this occupational polarization resulted in wage polarization? Here the data are a bit more ambiguous, but in general the answer seems to be yes. Wages have grown the most for those in professional, technical, and managerial occupations, perhaps because they have become more productive. Wages for those in administrative and manufacturing occupations have increased the least, most likely because the demand for labor in these occupations has shrunk faster than the supply of workers available to do the jobs. Finally, wage increases for those in occupations such as food preparation, personal care, and protective services fall between the other two. While the demand for labor in these occupations has been relatively strong, the supply of workers willing to do these jobs has grown faster. In part this is because many of those who were displaced from administrative and manufacturing occupations gravitated to these lower-paying jobs, mirroring what has happened as a result of the global outsourcing and offshoring of existing work. These jobs are most vulnerable to declines in unions and in the purchasing power of the minimum wage.

The Absence of Countervailing Forces

Management initiated these responses to globalization, deregulation, and financialization in attempts to achieve or increase competitiveness. It might be expected that to some extent the responses—or their effects—might have been countervailed by labor unions and/or by public policy-makers, but with rare exceptions they weren't. Throughout this period, labor unions were on the defensive and were basically fighting rearguard actions just to keep things from getting worse, while politicians, at least at the federal level, were locked in an ideological stalemate. We'll have more on these developments later, but here is a brief synopsis of what happened.

In 1981, President Ronald Reagan fired virtually all of the country's air traffic controllers for conducting an illegal strike. This action, coupled with the pressures on profits, emboldened many members of management to take a hard line with labor unions. In part, this involved efforts to disenfranchise existing unions and in part it was a matter of vigorously opposing any attempts unions made to gain new members. While other factors were in play, these tactics worked. In 1983, there were 18 million union members in the United States (constituting 20 percent of the labor force at the time). By 1996, this number had declined to 16 million (14 percent), and in 2017 it was down to 15 million (10.6 percent—6.4 percent in the private sector and 34 percent in the public sector). Meantime, strikes, long a key source of union power, lost their clout. While in the period 1950 to 1970, strikes generally paid off for union members in the form of larger pay increases, after 1980, this relationship no long held. In fact, it has been estimated that as much as 20 to 30 percent of the growing gap between productivity levels and workers' earnings since 1980 (shown in Figure 1.3) can be attributable to the flagging power of unions.

As for public policy during this time, little can be said. Although the ever-increasing earnings and wealth inequalities have been well known and understood for at least two decades, the U.S. government has taken virtually no action to stem the tide. It is fair to say that to date politicians and others responsible for the formation of public policy in the United States have essentiality been fiddling while the American Dream burned (apologies to Nero). It is little wonder that the country witnessed what essentially was a populist revolt that fueled Donald Trump's victory in the 2016 presidential election.

So this is path we have followed. Nearly four decades of managerial decision making under intense global competition, extensive deregulation, and extensive Wall Street scrutiny with few countervailing forces to stand in the way have brought us to the point where companies clearly are more innovative and agile and certainly more cost competitive than they were before. Consumers and the owners of capital have been the big winners, while middle-class and lower-level workers have been the clear losers. Wealth and income equality are at levels that have not been seen since the Great Depression, and thoughtful observers everywhere agree that the situation is unsustainable and has to change. The question, of course, is in what ways?

Where to from Here?

Any attempt to look ahead necessarily requires making some baseline assumptions about the important developments with respect to the major contributing factors discussed above—globalization, deregulation, and financialization and most of all automation or what is now popularly called *digitization*. We will touch briefly on the first three factors and then focus more intensively on the ways digital technologies might affect the future of work. A key point in all of this is that none of these forces are deterministic. They can be shaped by policy and by choices made at the company level and by individuals. Policies and choices can either contribute to or detract from building a new social contract.

The Forces That Will Be at Work and the Likely Implications

Globalization

The pace of globalization tends to ebb and flow in rather long cycles over time. As noted above, it flourished in the period between 1980 and 2008, but has slowed since then and is expected to continue to grow at a moderate pace for the foreseeable future.[12] There is no doubt that globalization has its benefits and beneficiaries, but it also has its costs and casualties. In the latest go-round, one clear group of casualties was workers on the middle and lower rungs of the income ladder. As these casualties accumulated and the resulting inequalities in earnings and wealth became clearer, nationalism and populism gained footholds that, in turn, are showing up in the form of increasing pressures on politicians to favor a range of protectionist policies, including the rejection of trade deals that fail to provide assistance for affected workers (e.g., the much-maligned Trans-Pacific Partnership). As the pace of world trade slows, global companies and companies that depend on them will feel the full effects in the forms of sluggish economic growth and tighter competition for a bigger share of an at best slightly expanding pie. This will intensify the current pull and tug between tight cost controls and innovation and agility in the marketplace to new heights and managements will struggle to find ways to accommodate both.

It is becoming increasingly clear that business and government leaders underestimated the gaps between who would win and who would lose as a result of globalization. Any effort to build a new social contract will need to address these gaps. One option is to slow or try to reverse globalization through trade barriers or tax policies that reward domestic production and job creation or that penalize firms that move work abroad. While some corrective measures may be sensible in order to limit currency manipulations or other ways of "dumping" imports into the United States at

[12] Ruchir Sharma, "When Borders Close," *New York Times*, November 12, 2016, accessed February 23, 2017, http://www.nytimes.com/2016/11/13/opinion/sunday/when-borders-close.html?r=0.

prices that do not reflect the true costs of production, these are not likely to bring back significant numbers of jobs that have already been lost abroad or significantly slow down further job losses in cases where lower-wage countries have a competitive advantage. Instead, countries like the United States will need to try to stay one step ahead by increasing investments in research and development and infrastructure that support our capacity to export while helping to create new high-quality jobs. One example of this approach is Manufacturing USA, a public-private network the federal government has created to sponsor research, development, and pilot testing of advanced manufacturing technologies in different sectors (see Box 3.1).

At the same time, increased resources will need to go into redeveloping both the labor market and communities in the regions that have most hurt by international trade. In the next chapter, we will stress the need for business leaders to focus less on holding down wages in order to compete with low-wage firms abroad and instead pursue high productivity and what we call high-road business strategies and employment practices.

Box 3.1. Manufacturing USA Initiatives

Manufacturing USA brings together industry, academia, and federal partners in a growing network of advanced manufacturing institutes, industries that promote the implementation of advanced manufacturing technologies. The goal is to increase U.S. manufacturing competitiveness and promote a robust and sustainable national manufacturing R&D infrastructure that will create high-quality manufacturing jobs and increase our global competitiveness. Investing in technologies such as information technology, biotechnology, and nanotechnology will support the creation of good jobs by helping U.S. manufacturers reduce costs, improve quality, and accelerate product development.

Source: https://www.manufacturingusa.com/.

Since its inception in 2011, the federal government has created twelve advanced manufacturing industry institutes that mix private and public funding with the expectation that this will grow to between 15 to 20 in future years (see Box 3.1). Each is funded with an initial grant from the federal government with matching contributions from private industry, participating universities and community colleges, and state governments.

Deregulation/Regulation

The communications and transportation industries have long since adjusted to deregulation and are not likely to see more than modest adjustments in the near future. Instead, the coming debates over industry regulations are likely to focus on the emerging trend of using independent contractors rather than regular employees. As we will discuss more fully in each of the following chapters, this debate is already well under way and the choices policy makers, technology developers, and employees and their potential labor representatives make will feature prominently in shaping the future of work and any new social contract. Stay tuned as we take these issues up in depth throughout the rest of this book.

Financialization

A growing number of and increasingly vociferous groups are critical of financialization and the short-term, shareholder value form of capitalism it favors. These critics have a strong hand when it comes to public opinion; one current survey showed that only 49 percent of the respondents believed that free enterprise is better than government at lifting people out of poverty, primarily because under capitalism "the rich get richer while the poor get poorer."[13] So the winds of reform are in the air. Still, financial firms, pension funds, mutual funds, and the like will remain economically and politically powerful. They are superbly skilled at lobbying against any legislative initiatives they perceive to be inconsistent with their interests. Thus, it seems likely that for the foreseeable future any major changes on this front will have to come from members of management who resist the pressure for short-term results and instead take the high road with respect to people and profits. Some may choose to take this path voluntarily, while others may bow to mounting public pressure or perhaps negotiate changes with labor unions or other agents that represent the interests of employees. How this might be done is the major focus of Chapter 6.

[13] Tim Montgomerie, "Has the World Lost Faith in Capitalism?" *Wall Street Journal*, November 6, 2015, accessed February 23, 2017, https://www.wsj.com/articles/has-the-world-lost-faith-in-capitalism-1446833869.

The bottom line is that it appears that both global corporations and even many smaller domestic firms are in for a future characterized by considerable volatility and stiff competition in the marketplace and by constant pressure from both the forces of financialization and advocates of change.

Digitization

As daunting as the foregoing developments are likely to be, digitization promises to pose even greater challenges on the employment front in the years ahead. Digitization is automation on steroids. Essentially it is the process of converting virtually all types of information—text, videos, music, photos, maps, speech, and data that are derived from sensors that can be embedded almost anywhere—into the ones and zeroes that computers use to work their magic. Ubiquitous digitization yields big data that opens up new worlds of possibilities not previously imagined, particularly when digitization and big data are combined with the exponentially increasing power of computers. Advances in manufacturing using 3D printing technologies that harness computer power to fabricate physical materials will also bring huge changes. As MIT researcher Neil Gershenfeld says, this technology allows anyone to make almost anything anywhere.[14] How these digital technologies will affect the world of work is one of the central debates of our time. The good news is that leaders as diverse as former president Barack Obama, business leaders in Europe (and maybe someday in the United States?), and technology developers and advocates agree that managing the development and use of these technologies better than the way than we did the last several decades of globalization is critical to building a new social contract.

[14] Neil Gershenfeld, "How to Make Almost Anything," *Foreign Affairs* 91 (November/December 2012): 42–57, accessed February 16, 2007, http://cba.mit.edu/docs/papers/12.09.FA.pdf.

Figure 3.2. Growing Voices on Digital Technologies and a New Social Contract

Diverse Voices	Their Key Message
Barack Obama in an interview with Scott Dadich, editor of *Wired* and Joi Ito, director of the MIT Media Lab[15]	The social compact has to accommodate these new technologies . . . or populist movements will see artificial intelligence as threatening to themselves and their community values.
European CEOs: Notes from a private discussion on technology and the future of work	We must manage the transition to advanced digital technologies better than the globalization process has been managed to date.
Tim O'Reilly, entrepreneur and technology expert[16]	One of the key social and economic questions that needs to be asked is whether network businesses (and other technology businesses) inevitably produce only a small core of high-paying jobs and a much larger network of lower-wage jobs, or whether this is the result of management choices and social policy.
Vivienne Ming, artificial intelligence expert and entrepreneur[17]	We need social institutions to keep pace and we need technologists to think about how their work will have a reverberating effect on society. As stakes for [the effects of technology] become higher and higher I think these should be collective decisions we make as a society because they have broad societal implications.

Historically, firms have tended to use automation in one of two ways. Substitution uses machines to replace most of the tasks associated with jobs or even to replace the jobs altogether, while augmentation attempts to use technology and talent in ways that combine the strengths of both to enhance both performance and the

[15] "President Obama on How Artificial Intelligence Will Affect the Future of Work," YouTube Video, accessed February 16, 2017, https://www.youtube.com/watch?v=DgL32wtgeXQ.

[16] Tim O'Reilly, *What's the Future of Work? Exploring the Economic Shift Led by Software and Connectedness* (San Francisco: O'Reilly Media, 2016).

[17] From Artificial Intelligence to Augmented Intelligence: A Conversation with Dr. Vivienne Ming, October 13, 2016, https://www.dropbox.com/sh/7t11gw46ox31smk/AABbJ3A4FWZTNikHCfLagg4La?dl=0.

quality of jobs. We have the same set of choices about how to use digitization at work. Already we see evidence of both approaches. In the past, robots were primarily used in manufacturing to take over tasks that were considered too taxing, too dangerous, or too mundane to be done by humans. Examples include heavy lifting, applying noxious substances such as glues and paint, and routine welding. These machines tended to be bulky and immobile, have limited functionality (i.e., they are programmed to do only one or at most a few tasks), lack dexterity, and be so big and heavy that they were dangerous to be around. Not surprisingly, armed with enhanced digitization and ever-increasing computer power, roboticists the world over are rapidly developing machines with wider applications that largely overcome these limitations. In some cases, these more flexible robots are being or will be used to eliminate work previously done by employees. Kroger, a large grocery chain, for example, has automated one of its warehouses to the point where robots organize the goods that come in from suppliers, making it possible to ship them out in ways that facilitate the stocking of store shelves. Although human workers unload and load the trucks, the intervening steps—inventorying, classifying, storing, and rearranging goods—are done without any human involvement.[18] Soon perhaps even our hamburgers may get the same treatment. A company called Momentum Machines has developed a robot capable of shaping burgers and grilling them to order, simultaneously toasting the buns, and slicing and adding accoutrements such as tomatoes, onions, and pickles to match customers' orders—again without any human involvement. Momentum's machine is capable of chunking out 360 hamburgers an hour. The machine is expensive and not always economically feasible given the wages paid in most fast-food restaurants today. But as Momentum's costs come down and prevailing wages go up, a tipping point may be reached where it might be wise to modify Momentum's machine to combine its capabilities with the skills of workers in an optimal way. This, however, is not the goal of the firm's founder. His goal is "not to make fast food workers more efficient . . . [but] to completely obviate them."[19] We take issue with this view of how technology should be used. This statement illustrates the point Vivienne Ming made in Figure 3.2: Technology is too important to be left to the technologists.

[18] Martin Ford, *Rise of the Robots: Technology and the Threat of a Jobless Future* (New York: Basic Books, 2015), 17.

[19] Ibid., 12.

In other cases, however, so-called collaborative robots that are designed to operate in tight spaces with few if any protective boundaries can stop on a dime if a person accidentally bumps into them. These robots are being used in warehouses and factories to augment the work of employees. Amazon, for instance, uses them in its newer warehouses to move goods around on about eight miles of conveyor belts, but unlike Kroger, Amazon has chosen to rely on workers to perform critical steps.[20] In these facilities, pallet-sized robots carry shelves of merchandise to strategically located stations, where workers have computers that help them identify the items they need. The workers then pluck the items off the shelves and scan them, after which a flashing light shows them which bins to place them in for shipping. In this case, the robots do the heavy lifting while also providing information that facilitates the quick and accurate movement of merchandise through the facility. Amazon has found that its robotized warehouses actually employ more workers than its older facilities, primarily because they are more efficient and are thus able to process a larger number of orders per day. Another example of augmentation comes from a Whirlpool plant in Marion, Ohio, that produces 18,000 clothes dryers a day.[21] It uses several collaborative robots to perform up to 40 different, largely repetitive tasks that include moving large drums, testing lint traps, and checking the backs of dryers (which workers have trouble reaching) for defects.[22] This frees up the workers to do the higher-skilled and more refined tasks such as mounting motors and installing, checking, and where necessary fixing the intricate wiring found in today's dryers. The idea is not to reduce the number of human workers but rather to improve productivity and enhance safety. One Whirlpool worker captured the essence of augmentation this way, "If I can get some help doing my job, I'm all for that. It's technology helping manpower [sic]—you can't beat that."[23]

AI—essentially the ability of machines to think and learn—has been around since the mid-1950s, but recently the research has made major strides in areas such as

[20] Laura Stevens, "How Amazon Gets Its Holiday Workers Up to Speed in Two Days," *Wall Street Journal,* November 28, 2016.

[21] Andrew Tangel, "Latest Robots Lend a Helping Arm at Factories," *Wall Street Journal,* November 9, 2016.

[22] Ibid.

[23] Ibid.

natural language processing; pattern recognition, problem-solving, and decision-making; learning; and even emotion detection. So far it seems that few companies are using this technology as a total substitute for labor and only a few are opting for pure augmentation; in most cases, there seems to be some combination of substitution and augmentation. A company called Lionbridge, for instance, uses an advanced translation technology to serve customers in multiple languages. The technology works with both written and oral communication, using machine learning to increasingly improve the quality of translations across industries and products. This facilitates more reliable interactions between non-English-speaking customers and English speaking customer service representatives. This reduces the need for translators and improves the quality of service customer service reps provide, since they are able to give more personalized and more accurate services to a greater range of customers. A firm by the name of Narrative Science now offers a service that automatically scans reams of company data (not only financial statements but also blog posts and the like), analyzes the information, and then writes reports that currently are composed well enough for internal (although not yet external) company use. Obviously, this eliminates the need for many analysts but it also frees up writers to do the more demanding and interesting tasks associated with preparing documents for public use (at least until the technology learns how to do these tasks as well).[24] Going one step further, a vendor known as HP Autonomy has developed a system that not only plows through thousands or even millions of legal documents to find appropriate precedents for legal cases but also suggests which ones might be most useful in any particular case. HP Autonomy claims that its system enables a single lawyer to accomplish the work it would take 500 lawyers and legal assistants to do the old-fashioned way, although there is no doubt that it also enhances the effectiveness of the highly-skilled lawyers who remain.[25] Still another approach is to use AI to teach machines how to read human emotions—joy, surprise, sadness, anger, confusion, and so on. For examples, it is now possible to install sensors in the steering wheels of trucks that tell drivers when they are too tired or stressed to stay on the road and alerts them to take a break. This is an example of pure augmentation.

[24] Geoff Colvin, *Humans Are Underrated: What High Achievers Know That Underachievers Never Will* (New York: Penguin, 2015), 20.

[25] Ibid., 17.

One of the leading forces in AI and cognitive computing is IBM's Watson. While Watson is known as champion game player (Jeopardy, Game of Thrones), it also applies a plethora of cognitive capabilities—natural language, pattern recognition, problem solving, decision-making, and learning—to help doctors do a better job of treating their patients, among other things.[26] Interestingly, an intricate collaboration between talent and technology is necessary for Watson to learn well enough to be trusted. In each area of medical specialty (oncology, radiology, endocrinology, etc.), the process starts with knowledge extraction: Watson scans mountains of medical information, finding common phrases and associating them. Then, knowledgeable physicians and scholars in each area look for errors and, if necessary, reweight the algorithms. The process is repeated until the experts are satisfied that Watson can produce correct diagnoses. Here is how Rob High, chief technology officer for IBM Watson, described what they are doing:

> We can actually set the system up . . . to identify the best treatment based on outcomes, based on the standard of care practices and clinical expertise, based on similarity of this patient to other patients. And all those can contribute to help Watson come back with an ordered list of potential treatments. But we're not trying to make the decision for the human . . . we're trying to do the research for the human so the human can then go make the decision better.[27]

The first stage of building the Watson AI system is a clear example of the workers giving wisdom to the machines, as the Japanese say. Once Watson is fully up to speed, the tables are turned and it is the machine that is giving wisdom to the workers. Then the workers take over again: A team of physicians, nurses, technicians, medical assistants, and patients can decide on the best course of treatment in each case and allocate the tasks of implementing, monitoring, and adjusting the plan as circumstances dictate. Once again this illustrates a key point: When ideas and knowledge about what technology and humans can do best across the full spectrum of the work to be done are integrated in order to solve a problem (how to best diagnose

[26] Clinton Leaf, "IBM Watson: Not So Elementary," *Fortune*, October 21, 2016, accessed February 23, 2017, http://fortune.com/david-kenny-ibm-watson/.

[27] Comments made at conference on Technology and the Future of Work. MIT15662T115-V014500, YouTube video, accessed February 16, 2017, htpps://www.youtube.com/watch?v=U-gQSxL78ks.

and treat a patient's illness) and when the people who have the biggest stake in solving the problem efficiently, effectively, and to the "customer" or patient's satisfaction are involved in the process, tasks can be redistributed among team members who use technology skillfully and perhaps even discover new uses or applications of a new technology. We believe that if this strategy is more widely used, AI and its human partners will find more uses for advancing technologies, use them more efficiently, create new work flows, and change the nature of work people do. The key, however, is not to let the technologists define the problem or the design parameters for the technology. Problem definition, information flows, work design, and development of new applications should be more broadly owned by the key stakeholders involved.

Many other examples could be cited. But given all these remarkable developments, with many more to come, what is the "so what" for employment? Some pundits are already asking if there will be any work left to do. The answer to this surely is yes. But how much and of what types is less clear. According to research conducted by the McKinsey Global Institute, only 5 percent of current occupations are likely to disappear entirely within the next decade, but there is a good chance that some 30 percent of the tasks currently included in about 60 percent of existing occupations will vanish or be changed significantly during that time.[28] The research done by MIT professor David Autor and his colleagues indicate that thus far the occupations that have been most affected by automation are those involving routine tasks. This trend is expected to continue. However, future developments in robotics, AI, and cognitive are likely to encroach on many previously untouched nonroutine or cognitive tasks that are common in technical, professional, and managerial occupations. At this point it seems that technological incursions into nonroutine or manual tasks will be slower in coming. This leaves only service-oriented jobs relatively untouched for now.

[28] Michael Chui, James Manyika, and Mehdi Miremadi, "Where Machines Could Replace Humans—And Where They Can't (Yet)," *McKinsey Quarterly*, July 2016, accessed February 23, 2017, http://www.mckinsey.com/business-functions/digital-mckinsey/our-insights/where-machines-could-replace-humans-and-where-they-cant-yet.

Recently, the Pew Research Center asked a sample of experts the following question: "Will emerging technologies displace more jobs than they create by 2025?"[29] Had the group been asked to comment on the historical experience, it is quite likely their responses would have been almost unanimous: Of course technology has eliminated many tasks and even jobs in the past, but it also has created many new tasks and jobs as well, so it has been a wash. When put in the role of forecasters, however, there was no unanimity. Instead, the group split right down the middle. Fifty-two percent said "no," while the other 48 percent said "yes." The optimists called on historical precedent. The comment by Professor Lawrence Katz of Harvard was typical; he said, "Over the long term, employment rates are fairly stable. . . . People always find things to do." The pessimists, however, were more convinced that things are different this time. Professor Eric Brynjolfsson of MIT was one of those; as he put it, "When all these science-fiction technologies are deployed, what will we need all the people for?" Even Professor Larry Summers, a well-known economist and a former president of Harvard who generally is not known for indecisiveness, was quoted as saying, "Until a few years ago, I didn't think this was a very complicated subject. I'm not so completely certain now."[30]

Notice that Pew's question asked for a finite answer; it asked "How will things turn out?" We suggest this is the wrong way to frame the question. We interpret the inconsistent and ambiguous answers to mean "it depends." It depends on who makes the critical choices about the problems we are trying to use technology to solve, who is involved in the design and implementation and further development of the technology, and how work processes are integrated and changed to make best use of the technology. That is, how things will turn out depends on the choices that current and future designers of technology make and on decision makers within companies who implement the technology. If the designers continue on the current path of thinking primarily in terms of eliminating human judgment and variation—recall the goal of Momentum Machine's founder to "obviate" the jobs of fast-food workers—then the pessimists may well be proven right. This is especially true if decision makers within companies opt for the elimination of labor costs through substitution instead of improving innovation, agility, and productivity through augmentation.

[29] Aaron Smith and Janna Anderson, "AI, Robotics, and the Future of Jobs," *Pew Research Center*, August 6, 2014, http://www.pewinternet.org/2014/08/06/future-of-jobs.

[30] Ibid.

But neither group has to follow these proclivities. They can choose to focus on the work to be done, especially on making improvements in the processes by which work is done and on using robotics and decision aids to eliminate some of the more arduous, unpleasant, boring, and dangerous tasks that people do. Simultaneously, they can choose to focus on changing how the remaining tasks are done to make them not only more efficient but more interesting and safer (as the Whirlpool and Watson examples illustrate). If this happens, the optimists in the Pew study will have been right on. However, it is clear that over time many tasks and some jobs will be eliminated and lost, another challenge that certainly will demand the collective attention of employers, the work force, educators, labor unions and other worker advocacy groups, and public policy makers in the years to come. As the European CEOs stated in Box 3.2, the lessons learned from globalization should not be lost as we cope with advancing technologies: There will be some winners (those whose jobs will be augmented or upskilled and those with the skills needed for new jobs yet to be created) and losers (those who will be displaced). Addressing the interests and needs of those in *both* groups, as Barack Obama noted, has to be part of a new social contract governing work.

Video Resources

The following videos covering topics in this chapter can be found at http://iwer.mit.edu/speak-up-for-work/:

Out of the Ashes: Starting Assumptions for a New Social Contract
Strategies for Competing in a Global Economy
Technology and the Future of Work: An Introduction
What Occupations are Growing: Interview with Heidi Shierholz
Technology and the Future of Work: Interview with Erik Brynjolfsson
How Will New Technology Influence the Workplace?
Interview with Vivienne Ming on Artificial Intelligence and the Future of Work
Interview with Neil Gershenfeld on Digital Fabrication and the Future of Work

4

How Business Leaders Can Contribute to a New Social Contract

This chapter looks at what business leaders can do to contribute to a new social contract at work. The first obligation is to assure that the business and employment strategies they use in their own companies deliver both solid financial results and high quality jobs and careers. The corollary obligation is to collaborate with other key stakeholders—leaders of labor and social justice organizations, educational institutions, the work force, and governments—to build an infrastructure of institutions, policies, and practices that reinforce and strengthen these strategies. In this chapter, we address the first issue in depth by discussing the essence of the preferred business strategies and the employment strategies that are most likely to deliver good results for both shareholders and employees. We also discuss the factors that encourage and discourage the implementation of these strategies and the actions business leaders can take to move things forward in a positive direction. We discuss these issues in the context of both standard employment relationships with a well-defined employer and employees who are covered under the full array of employment policy protections and the growing number of "fissured" arrangements involving domestic and/or global outsourcing of work or independent contracting in the so-called platform economy. In this chapter, we focus on actions individual employers can take. In Chapter 8, we discuss the joint actions business leaders need to take in collaboration with their peers, with others in civil society, and with government.

Our basic premise is that key decision makers in the world of business have *choices* about the employment strategies they pursue and that these choices have major consequences. Professor Zeynep Ton of MIT brings this point home clearly in the introduction of her book, *The Goods Jobs Strategy*:

> There are different ways to make money, I tell my students on the first day of the class I teach on operations for service industries. You can certainly succeed at the expense of your employees by offering bad jobs—jobs that pay low wages, provide scant benefits and erratic work schedules, and are designed in ways that make it hard for employees to

> perform well or find meaning or dignity in their work. You can even succeed at the expense of your customers; for example, by offering shoddy service. . . . Many people in the business world assume that bad jobs are necessary to keep costs down and prices low. But . . . [this] . . . is not a necessity, it is a choice. There are companies in business today that have made a different choice, which I call the good jobs strategy. . . . These companies—despite spending much more on labor than their competitors do in order to have a well-paid, well-trained, well-motivated workforce—enjoy great [financial] success.[1]

We readily accept that the primary duty of business leaders is to develop business strategies that increase the competitiveness of their companies and generate good financial returns for owners and shareholders. In this context, business owners face two challenges: finding employment strategies that not only support the attainment of these goals but also deliver high-quality experiences for the workers involved and finding ways to do so even as companies continue to morph into more diversified networks of complex work arrangements.

Why the High-Road Approach?

Figure 1.1, reproduced here as Figure 4.1, suggests that the basic choice business leaders face with respect to employment strategies is between what researchers have come to call a low-road strategy and a high-road strategy. Studies have shown that both strategies can produce favorable organizational outcomes, depending on the circumstances, but as Professor Ton makes clear, only the high-road strategy consistently delivers positive outcomes for employees as well.

[1] Zeynep Ton, *The Good Jobs Strategy* (New York: Houghton Mifflin, 2014), vii.

Figure 4.1. Options for How Businesses Can Compete

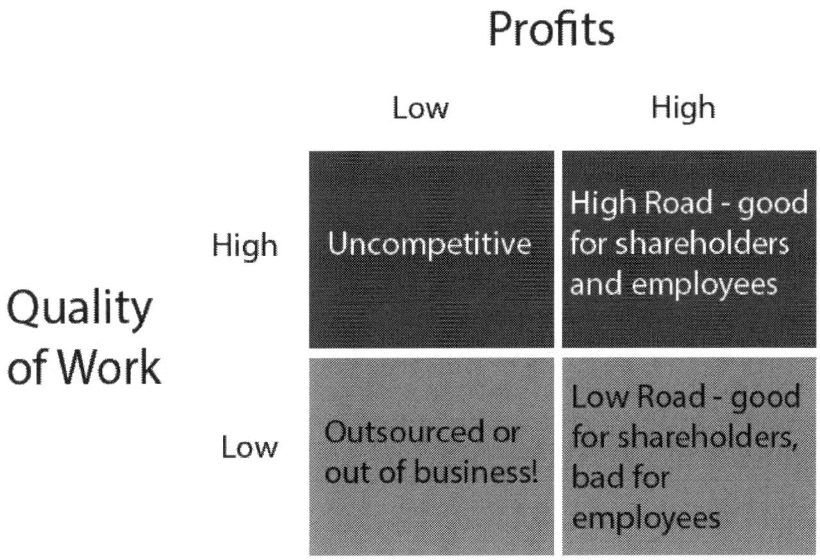

Figure 4.2 captures the essential features of these two strategies. Three points deserve emphasis. First, notice that the low-road strategy views labor primarily as a cost to be contained whereas the high-road strategy sees workers as assets or resources to be utilized to their full capacity. A second and related point is that the low-road strategy puts considerable emphasis on rules and policies designed to constrain and control employee behavior, while the high-road strategy takes a much more expansive view, seeking to enlist and enhance employee effort and discretion in the pursuit of company goals. These two strategies are sometimes referred to as being, respectively, control oriented and commitment oriented). And third, it cannot be emphasized too strongly that the high-road approach involves sets of mutually reinforcing policies and practices that must be developed and implemented as an integrated system. This means that while companies can start down this path in incremental steps, eventually it is necessary to go all in to achieve the best results. There are no shortcuts.

Figure 4.2. Key Elements in High-Road and Low-Road Business Strategies

	High-Road Strategies	**Low-Road Strategies**
View of Employees	Valuable resources	Costs
Selection of Employees	Careful; high standards	Casual; minimum standards
Job/Work Design	Multiple tasks	Very specialized
Skill Level and Scope	High, wide	Low, narrow
Training	Extensive; employee development	Limited; job specific
Job Autonomy/Discretion	High	Low
Hierarchy	Low differentiation between managers and workers; few layers	Rigid differentiation between managers and workers; many managerial layers
Wages	Relatively high wages; additional rewards linked to performance	Relatively low wages
Motivation	Intrinsic; focused on self-control	Extrinsic; focused on company control of workers
Employee Attachment to Company	Extensive	Minimal
Employee Voice	Informal and/or formal voice; options for engagement and representation	Little or no voice or engagement; representation suppressed

A comparison of the business and employment strategies of two well-known firms, Walmart and Costco, illustrates the differences between the low-road and high-road strategies.[2] Although both companies have had their ups and downs, overall both have been quite successful, although Walmart has struggled some recently. Between 2010 and 2015, for example, Costco's stock price went up 160 percent, while Walmart's increased by only 60 percent. Walmart's business strategy, as summarized in its familiar slogan "Everyday low prices, always," is unambiguous. Until very recently, the company chose to pursue this goal with a low-road employment strategy that had a laser-like focus on minimizing labor costs. Key features of this strategy included making minimal investments in employees, especially in terms of training and development, having tightly defined job descriptions, closely supervising employees, paying relatively low wages (retail associates started at around $9.00/hour with minimal benefits), and tolerating, if not encouraging, high levels of turnover (in order to replace more senior and thus relatively highly-paid employees with lower-paid ones—this helped keep Walmart's average hourly wage for in-store associates down to just a little over $13.00/hour), and avoiding unions at all costs.

Costco's business strategy differs slightly from Walmart's, since it generally focuses on higher-quality products and offers better service. Its employee strategy closely follows the high-road model, which generally is referred as a high-performance work system. Employees are carefully selected for their intellectual and social skills and they receive considerable training especially regarding product knowledge and customer service. Once trained, they are granted substantial autonomy in making decisions about and implementing improvements to improve their job performance and service to customers. For assuming this responsibility, they are relatively well paid (retail associates start at a little over $13.00/hour and receive generous benefits). In addition, because turnover is low, the company has many long-service employees (so Costco's average hourly wage is significantly higher than Walmart's: $20.89/hour). Although Costco doesn't encourage its employees to unionize, some have chosen to do so and where this is the case, union-management relations generally are good. While solid data are lacking, anecdotes emanating from Glassdoor and various public reports strongly suggest that Costco ranks well above Walmart on measures of employee satisfaction and that workers see it as a much better place to work. (In 2016, employees at Costco rated the company 4.0 on Glassdoor's 5-point scale and 81

[2] Wayne Cascio, "Decency Means More Than 'Always Low Prices': A Comparison of Costco to Wal-Mart's Sam's Club," *Academy of Management Perspectives* (August 2006): 25–28.

percent recommend it as a place to work, compared to a 3.2 rating and 66 percent recommendation rating at Walmart.)[3]

Retailing is a low-margin business. How is it, then, that Costco's high-road approach, which obviously involves considerable investments in employees and higher wages, can contribute to its business success? Part of the answer lies in the fact that its prices (and thus its revenue per square foot of retail space) are higher than Walmart's. But the more telling factor is that its work system is far superior to the low-road approach in terms of developing and utilizing the skills of employees and motivating them to not only work harder, but also smarter—to work together to be more innovative in developing suggestions for improving store operations and customer service. As a result, Costco enjoys considerably higher worker productivity than Walmart on commonly used measures in retailing. Costco's revenue per employee in 2016 hovered around $575,000 and its profit per employee was around $11,500; comparable figures at Walmart were $220,000 and $7,000, respectively. To return to our earlier point, there is no business that cannot benefit from having employees who work harder and smarter and as a result are more productive than those in the firms they compete with.

Research conducted over a two-decade span confirms that high-performance work systems have a performance edge. This is the case not only in large companies in such diverse industries as steel, autos, airlines, telecommunications, apparel, health care, semiconductors, and computers[4] but also in a number of smaller firms in manufacturing, health care, and retail.

Well, you might say, that's all well and good for Costco and a few other companies, but what about other the other firms? To what extent has the high-road approach diffused across the U.S. economy?

[3] "Costco Ratings," *Glassdoor*, accessed February 17, 2007, https://www.glassdoor.com/Reviews/costco-reviews-SRCH_KE0,6.htm; "Walmart Reviews," *Glassdoor*, accessed February 17, 2007, https://www.glassdoor.com/Reviews/Walmart-Reviews-E715.htm.

[4] Eileen Appelbaum, Jody Hoffer Gittell, and Carrie Leana, "High Performance Work Practices and Sustainable Economic Growth," Center for Economic and Policy Research, April 2001, accessed February 17, 2017, http://cepr.net/publications/reports/high-performance-work-practices-and-sustainable-economic-growth.

The Extent and Nature of Diffusion

Unfortunately, at this point, there is no clear answer to these questions. For one thing, diffusion is a difficult thing to measure—does a business have to use every element of the high-performance work system in an integrated way to be counted as a high-road company or might it be enough if most—or even just a few—of the elements are in place? Furthermore, many of the companies that choose to use this this approach don't necessarily use the label, and it is common to find companies that use the approach among some, but not all, employee groups. Not surprisingly, then, when we take a close look at the diffusion issue, several different patterns emerge.

It is clear, for example, that in many, if not most, companies, the employment strategies that apply to executives, managers, and selected professional employees are comprehensive and well-integrated high-performance work systems, although the term is seldom used in this context. These employees are carefully selected, given numerous training and development opportunities, granted considerable autonomy in carrying out their jobs, and of course are very well compensated in terms of pay, opportunities for incentives (such as bonuses and stock options), and benefits.

In many cases, high-performance work systems can be traced to charismatic founders who believed strongly in the high-road approach and stuck with it as their firms grew and prospered. Well-publicized examples include Market Basket, Costco, Google, W. L. Gore (the maker of Gore-Tex), Whole Foods, SAS, Morning Star (the world's largest processor of tomatoes), and a host of others.[5] A particularly interesting case is Southwest Airlines, which consistently rates at or near the top of its industry in terms of productivity, quality of service, and customer satisfaction and has been the most profitable U.S. airline for the past thirty years. Not surprisingly, the firm also has a highly satisfied and deeply committed work force and is regularly rated as one of the best places in America to work. The fact that it is also a highly unionized firm demonstrates that high-quality union-management relations can be another contributor to a successful high-road, high-performance work system strategy.[6]

A number of firms have recently taken incremental steps toward the high-road strategy. Many of these, especially in traditionally low-wage industries such as fast-food operations and retail establishments, involve improvements in pay and work

[5] Gary Hamel, *The Future of Management* (Boston: Harvard Business School Press, 2007).

[6] Jody Hoffer Gittell, *The Southwest Airlines Way* (New York: McGraw Hill, 2003).

scheduling procedures. McDonald's, for example, recently starting raising the starting pay in its company-owned (nonfranchised) stores, and IKEA has committed to use the "Living Wage Calculator" developed by MIT professor Amy Glassmeier as a way of calibrating pay to the cost of living in its various store locations.[7] At the extreme end of a high-road pay scale, Dan Price, the CEO of a company called Gravity Payments, recently made quite a stir by announcing that henceforth the minimum wage in his company would be $75,000 a year and that he would be taking a substantial cut in his own compensation to make this possible.[8] Gap, a leading retailer, has chosen to modify the computerized program it uses to schedule employees in its stores to provide more advance notice of work times and longer periods between shifts to allow employees ample spells for family and other personal pursuits, including sufficient rest. Otherwise, moves toward the high road tend to focus on providing employees with improved opportunities for education and development. In 2015, for example, Starbucks negotiated reduced tuition rates for online courses at Arizona State University and offered tuition reimbursements for all employees who lack college degrees. Several other firms—including Deloitte, UPS, General Mills, Exxon-Mobil, and Proctor and Gamble—have taken somewhat similar steps.

These are encouraging examples and perhaps portend more to come and a more complete adoption of the high-road practices at some of these firms. We will return to the question of how to manage these changes later in this chapter and bring home the difficulties and opportunities involved with a concrete case example. But first we should discuss some alternative company forms that might be more natural fits for high-road business and employment strategies.

Alternative Corporate Forms

A number of companies have chosen to build a high-road strategy directly into their charters. Some of these are benefit corporations, whose charters commit managers to consider social and environmental outcomes in their decision-making, in addition to

[7] Dave Jamieson, "Ikea's Minimum Wage Hike Was So Successful, It's Raising Wages Again," *Huffington Post*, June 24, 2015, http://www.huffingtonpost.com/2015/06/24/ikea-minimum-wage_n_7648804.html.

[8] "Seattle CEO to Pay Employees $70,000 Minimum Wage," *USA Today*, April 15, 2015, http://www.usatoday.com/story/news/nation-now/2015/04/15/seattle-company-70000-minimum-wage/25810099/.

financial outcomes. Benefit corporations are chartered in thirty-one states and the District of Columbia.[9] Patagonia is a well-known benefit corporation.[10]

Another well-known corporate form that professes to value both shareholders and employee owners are what are known as employee stock ownerships, or ESOPS. In the United States there are currently over 10,000 companies that are ESOPS. The over 12 million employees of ESOPs own a significant (in some cases a majority) share of the company they work for.[11] Britain also has a long history of promoting ESOPs and currently has over 5,000 firms with significant employee ownership.[12] In both countries, some of these are simply done for tax-saving reasons. Experience shows that few of these companies reap any of the potential mutual benefits of employee ownership. Those that do benefit from employee ownership work hard to embed a culture of ownership among employees by giving them a voice in how they do their work and how operations can be improved and compensate them with both competitive wages and the prospect for long-term income growth if the firm prospers.[13] One of the best known and longest-lasting retail firms with broad-based employee ownership is the John Lewis Partnership in Britain. Box 4.1 reports how authors of a recent study describe this firm.

[9] "What Is a Benefit Corporation?" *Benefit Corporation*, accessed 24 February 2017, http://benefitcorp.net/what-is-a-benefit-corporation; "State by State Status of Legislation," *Benefit Corporation*, accessed 24 February 2017, http://benefitcorp.net/policymakers/state-by-state-status.

[10] Mat McDermott, "Patagonia Becomes a California Benefit Corporation," treehugger, January 3, 2013, accessed February 17, 2017, http://www.treehugger.com/corporate-responsibility/patagonia-becomes-california-benefit-corporation.html.

[11] Joseph R. Blasi, Richard B. Freeman, and Douglas L. Kruse, *The Citizen's Share: Putting Ownership Back into Democracy*. (New Haven, CT: Yale University Press, 2014).

[12] Graeme Salaman and John Storey, *A Better Way of Doing Business? Lessons from the John Lewis Partnership* (London: Oxford University Press, 2016), vii.

[13] Blasi, Freeman, and Kruse, *The Citizen's Share*.

Box 4.1. The John Lewis Partnership (JLP)

As a revered UK institution and a successful business, the JLP merits our attention. . . . It is extraordinary in terms of its distinctive features, its longevity, its size, its continuing success, its popularity with staff, customers, and the wider public. Longevity: the first shop was opened 150 years ago and the Partnership was formed nearly 100 years ago; size: the 2015 Annual Report and Accounts show 10 billion [British pounds] in sales and over 93,800 partners. The Partnership enjoys increasing appeal to politicians and management-writers for whom it represents a much-lauded, admirable, and moral alternative to the conventional form of business organization. Most of all, it's extraordinary because, while other firms find ways to reduce the number of staff or reduce their rewards and margins to suppliers, the JLP remains committed not to maximizing shareholder value (and senior management bonuses), but to partners' happiness.

Source: Graeme Salaman and John Storey, *A Better Way of Doing Business? Lessons from the John Lewis Partnership* (London: Oxford University Press, 2016), vii.

Cooperatives are another corporate form. The largest in the world is the Mondragon Group in the Basque region of Spain, a large worker-owned industrial cooperative formed in the 1950s by a Jesuit priest that has grown to employ about 60,000 workers in Spain and another 15,000 in other countries. IKEA is another well-known cooperative. The firm, which is headquartered in Sweden, has stores throughout the world. Novo Nordis (Denmark), Tata (India), and Natura (Brazil) are other international examples. Closer to home is our favorite professional football team, the Green Bay Packers. The Packers are owned by community residents who bought nontradable shares issued by the team in the 1920s and again recently when it needed to raise revenue to expand its stadium, Lambeau Field. A signal advantage of the cooperative form of organization is that it cannot be sold or become the victim of a hostile takeover by a financial investor seeking short-term gains. Stability and continuity and presumably a longer-term view are built into this organizational form. How else could Green Bay, Wisconsin, a city of about 100,000 people, have kept a professional, often a very good, and sometimes a championship football team for nearly a century?

Notwithstanding these examples of publicly traded firms and firms that feature alternative corporate forms, it should be clear at this point that the high road is the one less traveled. Why is this so if it is at least as good as, and often better than, the low-road approach in generating favorable business results and consistently better than the low-road approach in producing favorable outcomes for employees? Why haven't more firms adopted it? Well, it turns out that many factors influence this decision, both pro and con.

Factors That Foster and Hinder Diffusion

At the start, it is necessary to acknowledge that we lack solid evidence as to why some executives and managers choose to take the high road while others do not, although we can make inferences based on published materials and on our own personal experiences. In this spirit, we suggest that the decision to adopt and thus diffuse the high-road approach across businesses often comes down to the interplay between the fundamental beliefs of those who choose and shape the employment strategies of companies as well as the relative strength of various pressures on them to favor one strategy over the other.

Fundamental Beliefs of Decision Makers

Two fundamental beliefs of key decision makers influence their choice of employment strategies significantly. One is their view about the basic purpose of the firm and the other has to do with their assumptions about employees' attitudes toward work.

The Purpose of the Firm

What is the purpose of the firm? It turns out that this simple question masks a highly controversial issue that centers on two rather different perspectives that are often labeled the shareholder and stakeholder views. Those who favor the shareholder view take the stance that the business of business is business; that is, the purpose of the firm is to maximize profits and provide high returns to those who have put their capital at risk. Those who favor what is known as the stakeholder view take a broader perspective, noting that firms are social institutions and that in addition to owners there are many others—employees, customers, suppliers, communities, etc.—with skin in the game who thus have an equal right to have their interests considered. In the current context, employees are the stakeholders of interest. The question, then, boils down to this: Do firms exist to benefit only those with financial equity at stake or should they also be held accountable for ensuring that employees receive fair returns (financial and otherwise) for the sweat equity they put in? (Notice that this typically isn't an issue in firms with alternative forms of ownership, such as benefit corporations, employee-owned firms, and cooperatives.)

For the most part, the interests of owners have reigned supreme over the decades. During the period immediately following World War II, however, the stakeholder view took hold (although it wasn't until later that the term came into general use). As noted in Chapter 2, conditions were favorable for this development—U.S. business was dominant worldwide, oligopolies prevailed in most major industries, investors were largely passive, and labor unions were powerful. In addition, a new generation of business leadership had come to the fore. Many were first-generation college graduates from working-class backgrounds. This, along with the facts that this generation grew up during the Great Depression and many had served in the military during World War II, appears to have had a profound effect on their views. While they were hardly indifferent to making profits, they also had a deep sense of community and social responsibility and were conditioned to be concerned about balancing the interests of multiple stakeholders, including their employees. Box 4.2 profiles one prominent and in many ways representative member of this cohort, Peter McColough, who went on to become CEO of Xerox Corporation.

Box 4.2. Peter McColough, Harvard '49, Xerox CEO

Peter McColough was the CEO of Xerox Corporation from 1968 to 1982. Perhaps as much as anyone, McColough exemplified the type of leadership that supported and reinforced the postwar social contract. He played an active leadership role in community affairs in Rochester, New York, where the company was headquartered, and in national political and government advisory roles; he built strong and positive relationships with the union that represented Xerox manufacturing workers; he was ahead of other companies in initiating total quality and employee involvement processes in partnership with the union in the early 1980s; and, perhaps most important, as early as 1968 he initiated an affirmative action program that made it possible for women and minorities to rise to a wide range of executive leadership positions at Xerox. It is no accident that in 2002 Ann Mulcahy was named CEO of Xerox and that in 2009 she was succeeded by Ursula Burns, the first African American female CEO of a Fortune 500 firm.

In the early to mid-1980s, as noted in Chapter 2, things began to change. As profits became increasingly difficult to come by, shareholders and a number of aggressive financiers began to put considerable pressure on publicly traded firms, demanding increasing short-term profits and higher stock prices. Not coincidentally, this came just as labor unions were rapidly losing much of their power. And slowly but surely a new generation of business leaders came to the fore. These people were advocates of Milton Friedman's famous dictum: "There is one and only one social responsibility of business—to use its resources and engage in activities designed to increase its profits."[14] To reinforce this dictum, business school theorists invoked "agency theory," which advocated heavy incentivization of executives using stock options as a means of aligning their interests with those of the other owners.[15] At the same time, a new breed of executive compensation consultants were learning how to use the so-called Black-Scholes formula to put a monetary value on stock options, thus making it easier to hand them out.[16] The results weren't always pretty. Some executives were moved to engage in a variety of financial manipulations that ranged from buying back their own company's stock to outright fraud to drive up the value of their options. Nonetheless, the shareholder maximizing view (some called it the "financialization" of a firm) became deeply embedded in finance and economics teaching in business schools and has prevailed ever since.

Recently, however, the view that firms should prioritize maximizing short-term shareholder value has come under fire for being a major contributor to the growth in

[14] Milton Friedman, "The Social Responsibility of Business Is to Increase Profits," *New York Times Magazine*, September 13, 1970, http://query.nytimes.com/mem/archive-free/pdf?res=9E05E0DA153CE531A15750C1A96F9C946190D6CF.

[15] Michael C. Jensen and William H. Meckling, "Theory of the Firm: Managerial Behavior, Agency Costs and Ownership Structure," *Journal of Financial Economics* 3, no. 4 (1976): 305–360.

[16] Fischer Black and Myron Scholes, "The Pricing of Options and Corporate Liabilities," *Journal of Political Economy* 81, no. 3 (1973): 637–654; Robert C. Merton, "Theory of Rational Option Pricing," *Bell Journal of Economics and Management Science* 4, no. 1 (1971): 141–183.

income and wealth inequality, among other social ills. Indeed, many have come to question the virtues of this form of capitalism.[17] So, apparently, do a growing number of investors. Among the fastest-growing investment funds are the so called environmental, social, and corporate governance funds that invest in environmentally and socially conscious firms. Although they are still small by Wall Street standards, they attracted an estimated $8 trillion in investments between 2010 and 2016.[18] Clearly there is a taste for investing in firms that both provide a competitive rate of return and meet investors' expectations about protecting the environment and providing good employment conditions.

Whether the key decision makers in a company subscribe to the shareholder or the stakeholder view clearly matters when they make decisions about employment strategies. This is particularly true when it is considered in conjunction with their prevailing assumptions about employees' attitudes toward work.

Dominant Mental Models Regarding Employees

We use the term mental models to describe the assumptions of key decision makers in a company about employees' attitudes toward work. These views shape their decisions about the design of employment strategies. Many years ago, MIT professor Douglas McGregor provided a useful framework for understanding this that is still relevant today. In his well-known book *The Human Side of Enterprise,* Professor McGregor differentiated between two quite different mental models that he had observed in his research and consulting work.

One he labeled Theory X. Managers who subscribe to this view assume that workers generally are uninterested in company goals or affairs, dislike work and lack personal motivation, and prefer to avoid responsibility on the job. McGregor suggested that managers who base their decisions on these assumptions tend to

[17] For a range of critiques, see Sanford M. Jacoby, *The Embedded Corporation: Corporate Governance and Employment Relations in Japan and the United States* (Princeton, NJ: Princeton University Press, 2004); William Lazonick, *Sustainable Prosperity in the New Economy* (Kalamazoo, MI: W. E. Upjohn Institute for Employment Research, 2009); Eileen Appelbaum and Rosemary Batt, *Private Equity at Work: When Wall Street Manages Main Street* (New York: Russell Sage Foundation, 2014).

[18] Randall Smith, "Investors Sharpen Focus on Social and Environmental Risks to Stocks," *New York Times*, December 15, 2016. http://www.nytimes.com/2016/12/14/business/dealbook/investors-social-environmental-corporate-governance.html?ref=business&_r=0.

choose an employment strategy that favors little investment in employees, closely circumscribed rules and tight job descriptions, close (although not necessarily "harsh" supervision), and a mixture of inducements and penalties to encourage compliance with instructions and performance standards. McGregor labeled his second mental model Theory Y. In sharp contrast to theory X, it assumes that workers tend to identify closely with their organizations, like to work and welcome challenges on the job, and are internally driven to do well and better themselves through their careers. Managers who subscribe to Theory Y are inclined toward an employment strategy that favors extensive investments in employees, few guidelines, general job descriptions that allow considerable autonomy on the job, minimal supervision, and a reliance on self-motivation coupled with fair rewards that are commensurate with the contributions employees make.

Professor McGregor's work made two important contributions to discussions of employment strategies. First, his elucidation of Theories X and Y employment strategies were clear precursors to the more recently developed notions of low- and high-road (control and commitment) strategies described above. But of more direct relevance to the current discussion is the close causal connection he made—and many others have observed—between company decision makers' prevailing mental models about human nature and the employment strategies they choose to pursue. A Theory Y mental model is more consistent with a multi-stakeholder view of a firm. Most employees quickly see the inconsistencies between managers who espouse a Theory Y perspective but behave as if only short-term shareholder value is all that really counts in the ways they run the business. Or, as was brought home so vividly in the case of Market Basket, employees who perceive that owners or top executives are trying to change their company strategies and culture from a multi-stakeholder, Theory Y perspective might just fight back and gain support from customers and the public (see Box 4.3).

Box 4.3. The 2014 Revolt at Market Basket

Imagine high-level executives, store managers, clerks, and warehouse workers standing outside their stores side by side for a month demanding their CEO be reinstated and the business model that made the company thrive be maintained. And imagine their customer base cheering them while they had to shop elsewhere at considerable inconvenience and expense. That is exactly what happened this summer at Market Basket, a highly successful New England family owned grocery chain with 71 stores and 25,000 employees. It is clearly the biggest labor story of the year, and if it emboldens others to speak out for similar workplace causes, it may turn out to be the most important workplace event to come along so far in this century.

This broad-based revolt (aka strike) defied all traditional doctrines in labor-management relations, labor law, and corporate governance. It was the outgrowth of a long-standing feud within the owners, the Demoulas family, in which two cousins (Arthur T. and Arthur S.) vied for control of the business.

For years, Arthur T. had led Market Basket to high profits with a business model that provided consumers with low prices and good-quality service by building a highly productive, well-paid, and loyal workforce. But when Arthur S. gained control in June, he fired Arthur T., replaced him with new co-CEOs of his choosing, and began pursuing options to increase the flow of cash to family owners.

Employees demanded that Arthur T. be reinstated and that the business model they built together be restored. They organized rallies that attracted as many as 10,000 workers, customers, and community supporters. They used a "Save Market Basket" Facebook page to spread their message and maintain solidarity across the ranks. At one point, 68 of the 71 store managers signed a statement saying they would not work for anyone but Arthur T. Customers offered countless testimonials about the low prices and good service they were missing and documented the increased costs they incurred in shopping elsewhere by taping their sales receipts to the windows of their local Market Basket stores.

Unlike so many recent labor battles, this one ended happily. After weeks of negotiations, with a strong push for agreement by governors Deval Patrick of Massachusetts and Maggie Hassan of New Hampshire, the board of directors agreed to sell the company to Arthur T. and allow him to lead employees back to work and customers back to their beloved stores.

Source: Thomas A. Kochan, "What the Market Basket Deal Says about American Workers," Fortune.com, August 28, 2014, http://fortune.com/2014/08/28/ what-the-market-basket-deal-says-about-american-workers/.

The bottom line is this: a wider adoption of the stakeholder view of the firm and the Theory Y view of employees would go a long way toward convincing more employers to adopt the high-road approach to managing employees.

The Pull and Push of Outside Pressures

Pressures from outside groups also make a difference. As noted above, unions constituted one of the strongest forces that pushed firms to adopt high road strategies during the years of the postwar social contract. As early as the 1960s, scholars such as Sumner Slichter and his colleagues at the Harvard Business School noted that unions exerted a "shock effect" on management.[19]By this they meant that as unions use collective bargaining to achieve incremental wage increases, managers are forced to search for productivity increases to offset the high wage costs. As managers did this during the post-war period, they sustained the tandem increases in wages and productivity that we illustrated in Figure 1.3. The decline in union membership since the early 1980s has reduced this incentive. Recent estimates suggest that the decline in the bargaining power of workers accounts for between 20 and 30 percent of the rise in income inequality in the American economy since then.[20]

As union power declined, the power of Wall Street investment analysts increased as a force that promoted low-road strategies. CEOs focused on meeting investment analysts' expectations regarding earnings and revenue growth. Few analysts either understood or cared about what it would take to compete with high-road strategies. As one CEO once said in an off-the-record conversation, "When I talked about what we were doing to grow revenues and control costs, the analysts typed away on their computers and followed up with questions. When I tried to talk

[19] Sumner Slichter, James Healy, and E. Robert Livernash, *The Impact of Collective Bargaining on Management* (Washington, DC: The Brookings Institution, 1960).

[20] Bruce Western and Joseph Rosenfeld, "Unions, Norms, and the Rise in U.S. Wage Inequality," *American Sociological Review*76, no. 4) (2011): 513–537.

about how we were investing in our employees, their computers stopped, the eyes of those in the room glazed over, and they lost interest."

Government policies also played an important role. In the 1980s, key industries such as trucking, communications, and airlines were deregulated. The purchasing power of the federal minimum wage fell by over 25 percent between its peak in 1968 to today. The enforcement staff and budgets of key Labor Department agencies such as those that enforce wages and hours laws and safety and health regulations shrank even as the size of the work force and the number of workplaces increased. Congress tried to immobilize the National Labor Relations Board by holding up appointments of key officials, threatening to eviscerate its budget, and threatening to block implementation of update rules the agency issued to better enforce labor laws. Thus, there was little government pressure to raise the floor on the labor practices low-road firms favored.

Business schools reinforced the financialization of the economy and the focus on shareholder value as the underlying purpose of companies. Labor relations courses disappeared from course offerings at business schools and human resource management faculty were slow to take up teaching about the differences between high-road and low road strategies and employment practices. Just as was the case inside firms, finance came to dominate culture at most business schools.[21]

How to Transform a Low-Road Firm to a High-Road Firm: The Case of Walmart

What if you are an executive who wants to transform your organization from a low-road to a high-road firm? How difficult is this? How long does it take? How might you go about managing the change process? Let's use an extreme case to illustrate what's involved. Earlier in this chapter we described Walmart as the prototypical low-road firm. But recently it has announced its interest in changing and has taken some initial steps to do so.[22]

[21] Rakesh Khurana, *From Higher Aims to Hired Hands: The Social Transformation of American Business Schools and the Unfulfilled Promise of Management as a Profession* (Princeton, NJ: Princeton University Press, 2010).

[22] Much of the account of Walmart's actions that follows is taken from Neil Irwin, "How Did Walmart get Cleaner Stores and Higher Sales? It Paid Its People More," *New York Times*, October 15, 2016,

Evidence was accumulating to show that the company's low-road employment strategy was not producing the desired results. According to insiders, the company had come to be viewed as "a last-ditch option for employment" and was not attracting strong job candidates. Walmart had reduced the number of its employees by about 7 percent over the years, while the overall square footage available for sales had increased by 13 percent. This equation stretched employees thin. Consequently, the percentage of stores that were meeting their customer service goals had fallen to 16 percent. There were mounting customer complaints about dirty restrooms, empty shelves, endless checkout lines, and a lack of available employees to provide information and help for shoppers. Sales were on a downward slide, and in 2015 the company's overall revenue fell for the first time in its history. Profits were slipping as well, and the company's shareholders were not pleased. Clearly, a situation is ripe for change when shareholders, employees, and customers are all disgruntled. Walmart had also experienced increased pressures from employees and unions trying to organize or protest working conditions. The company had paid out nearly $1 billion to settle class-action lawsuits over violations of overtime rules and altering time-clock records to avoid paying for required meal and rest breaks. To top it off, a growing number of states were raising their minimum wage above the federal requirement; this put more pressure on the company to raise its entry-level wages. Pressures for change were mounting from many outside sources.

In response, Walmart announced in February 2015 that it intended to raise starting wages to at least one dollar above the required state or federal minimum. Unfortunately, Wall Street was not impressed. Between February 2015 and October 2016, Walmart's stock price declined 25 percent while the average for other retailers increased by 20 percent.

A key lesson of this chapter—that one cannot just change one employment practice such as wage levels to get the promised benefits of a high-road strategy—was not lost on Walmart executives. In October 2016, the *Wall Street Journal* reported that Walmart's executive team was also increasing its investments in training by $2.7 billion. In addition, according to Judith McKenna, the company's chief operating officer, Walmart was planning to take other actions to "not just fix one part, but get all of these things moving together". By "these things," she was referring to continuing to

http://www.nytimes.com/2016/10/16/upshot/how-did-walmart-get-cleaner-stores-and-higher-sales-it-paid-its-people-more.html?_r=0.

increase entry-level wages to $10 per hour and wages for department managers from $12 to $15 per hour, building 200 new training centers to foster employee development, and putting a new emphasis on career development at the store level by making it clearer to associates how they could advance to managerial positions and by using the new training centers to help employees develop the skills they needed to get ahead.

These changes are being phased in over time across Walmart's many stores. It is being done with great fanfare, so favorable results are already emerging. Store managers, for example, are finding a marked improvement in the quality of recruits. As one said, "We're attracting a different kind of associate. We get more people coming in who want a career instead of a job."[23] The percentage of stores that are meeting their customer service goals has jumped to 75 percent (from 16 percent). Sales are up 1.6 percent, a figure that might seem modest until it is compared with the industry average for the same period (0.4 percent). (Interestingly, some of these increased sales are coming from Walmart's own employees, who now can afford to shop there.)

The news is not all rosy, however. The *Wall Street Journal* has reported that Walmart's operating income was down 6 percent in the quarter ending September 30, 2016, and that the company's stock was still underperforming, in part because of the increased costs associated with these investments. Analysts are not convinced that the new approach will pay off anytime soon.

Moreover, Walmart has yet to address a number of other elements in a comprehensive high-road strategy. No efforts are yet under way to change the job structures and work processes to give employees the flexibility, knowledge, and discretion they need to solve customer problems. The company still needs to make changes that would enable its employees to better track and manage inventory. And, as we will discuss in Chapter 6, the biggest ideological or "mental model" hurdle of all for Walmart lies in its continued strong resistance to allowing employees to have their own voice in shaping company policies and workplace practices and the terms and conditions of their employment. Walmart continues to fight any form of collective action by its employees.

Changing a gigantic and widespread organization like Walmart is a daunting task. Some believe that picking one or a small number of locations and doing a

[23] Ibid.

wholesale transformation as a test or demonstration case would be a better strategy than trying to implement change gradually across an entire corporation. This is often called a "greenfield" strategy (i.e., start with a green field, build a new store there, hire a completely new work force, and don't carry over any of the management traditions from other locations). Then if the greenfield site is successful, a company can use it as a learning laboratory for changing other locations.

Walmart's experience to date is predictable and in line with the well-known "worse before better"[24] phenomenon that most companies have to go through as they make changes. That is, the costs are incurred and are visible to outside analysts long before the changes and investments begin to pay off in equally visible ways. How long it takes is uncertain, and this uncertainty and the difficulties of quantifying the benefits in ways Wall Street analysts understand or appreciate make it challenging for advocates of the changes to stay the course and maintain the support of skeptics within the managerial ranks.

Thus, Walmart faces many unanswered questions. Are the changes the company has made responsible for the favorable results obtained so far or does the credit go to the fact that changes are coming from the increased attention being paid to employees rather than to the specifics of the new approach? Do these changes represent the first steps toward a full-blown and sustained move toward a high-road employment strategy? Will Walmart abandon its ideological resistance to worker voice and representation? Will the advocates for change inside Walmart fall victim to the "worse before better" problem during the next economic downturn or with a change in top management? Alternatively, if Walmart continues on the path to the high road, will the company someday surpass Costco and its other competitors in terms of productivity and profitability and truly come to be seen as a great place to work? We shall see. Meantime, a "new Walmart" is certainly an intriguing possibility that is strongly to be encouraged as an example for other firms in its industry (and perhaps in related industries as well) and as an interesting experiment from which there will be much to learn.

[24] Nelson P. Repenning and John D. Sterman, "Nobody Ever Gets Credit for Fixing Problems That Never Happened: Fostering and Sustaining Process Improvement," *California Management Review* 43 (Summer 2001): 64–88, http://web.mit.edu/nelsonr/www/Repenning=Sterman_CMR_su01_.pdf.

Start-Ups and Platform Businesses

So far we have focused on large, well-established businesses. However, it is well known that new firms—start-ups—have historically served as an important source of job creation. Are start-ups following low-road or high-road strategies? Does this distinction even fit, given the increasing number of start-ups that are adopting "platform" forms of organization, or entities that use advances in digital technologies to link buyers and sellers of goods and services more directly without building large organizations? What role do these entrepreneurs and business leaders play in building (or resisting) a new social contract? We take up these questions now.

Say you are a young aspiring entrepreneur who wants to turn your great idea into a successful business. You also want to build a sustainable company, perhaps in both an ecological and a social sense. But you know that at least in the early years of your new venture, money will be very scarce and it will be tough to attract the talented employees you will need and pay them a good salary. What do you do? When asked this question, the advice of Scott Stern, one of our leading experts at MIT on start-ups, was "pick your investors carefully." Consider the example of one start-up that followed Scott's advice, Farmgirl Flowers.

Farmgirl Flowers founder Christina Stembel is committed to both developing a creative and successful online business that delivers flowers and to providing good jobs for her team members.[25] She faced an initial decision: Should they hire employees to do the work or was it better to go with independent contractors or free agents? Venture capitalists who are eager to provide much-needed capital often urge new entrepreneurs to use the latter strategy. Stembel decided to reject the advice of venture capitalists, hire employees, and work on developing high-road employment models.

But many others are following the alternative path. Indeed, many venture capitalists serve the function for start-ups that investment analysts serve for large corporations. They put considerable pressure on entrepreneurs to increase the valuation of the company in a short time and to hold down staffing costs by using outside contractors rather than hiring full-time employees whenever possible. Moreover, start-ups that begin with few resources struggle to provide high-quality jobs (competitive salaries with benefits and good learning and development

[25] https://farmgirlflowers.com/meet-the-farmgirl.

opportunities), except perhaps for the original founders and their top-level associates.[26] And how firms start often has long term consequences. A well-known study of Silicon Valley start-ups demonstrated that the employment strategies company founders implement tend to persist as a company ages, often even after the original founders have left.[27]

The good news is that there seems to be a growing number of young entrepreneurs who are building environmentally and socially sustainable organizations that serve useful social purposes. And a number of organizations are working to help them do so. Investors' Circle, an initiative supported by the Hitachi Foundation, works with young entrepreneurs who combine a passion for addressing some important social or environmental issue with the desire to build a profitable and sustainable new firm. It creates spaces for entrepreneurs to meet with investors looking for promising young firms with strong social missions.[28] They are in effect creating incubators and mentoring processes modeled after successful high-tech incubators. MIT is also encouraging more long-term and inclusive innovation initiatives. The Initiative on the Digital Economy sponsors an "inclusive innovation" contest each year to recognize (and reward with cash prizes) start-ups that address critical social issues and "create economic opportunity in a digital age."[29] In 2016, MIT announced its launch of The Engine, a new plan to support an incubator for start-ups that require a longer time horizon than most venture capitalists are willing to tolerate. The Engine will focus on building companies grounded in scientific and engineering innovations.[30] Whether these efforts will be successful in countering the short-term strategies of venture capitalists that focus on cost control remains to be seen. The point here is new entrepreneurs have choices about how to build their businesses and

[26] Adam Seth Litwin and Philip H. Phan, "Quantity over Quality: Reexamining the Relationship between Entrepreneurship and Job Creation," *ILR Review* 66 (July 2013): 833–873.

[27] James N. Baron, Michael T. Hannan, and M. Diane Burton, "Building the Iron Cage: Determinants of Managerial Intensity in the Early Years of Organizations," *American Sociological Review* 64 (August 1999): 527–547.

[28] http://www.investorscircle.net/.

[29] https://www.mitinclusiveinnovation.com/.

[30] Rob Matheson, "MIT Launches New Venture for World-Changing Entrepreneurs," *MIT News*, October 26, 2016, accessed February 17, 2017, http://news.mit.edu/mit-announces-the-engine-for-entrepreneurs-1026.

that these choices have predictable and long-term consequences for multiple stakeholders who help support and depend on the company's success.

Platform Technologies and Organizations

One of the most visible and potentially innovative trends in modern business is the growth of so-called platform businesses that use advances in digital technologies to connect customers and suppliers of goods and services without building large companies. Many of these "disruptive" (in the sense that they are disrupting the way companies provide these services) business models are achieving high levels of market capitalization (i.e., their estimated value if they were to be sold) without creating many jobs, or at least many good jobs. Figure 4.3 illustrates this with examples of some well-known platform companies. Consider one included in the figure: Netflix, the provider of online movie rentals. In 2006, Blockbuster, a previous supplier of movie rentals with retail outlets scattered across the country, had about the same level of sales as Netflix has today. Blockbuster employed 60,000 people that year. Within a decade, as Netflix and other platform movie providers took hold of the market, Blockbuster disappeared (all its stores were closed by 2013). Today, Netflix employs only 3,700 people, yet it has achieved the same revenue and nearly ten times the $5 billion market valuation Blockbuster achieved at its peak.[31]

[31] Gerald Davis, *The Vanishing American Corporation* (San Francisco: Barrett-Koehler, 2016); Christopher Harress, "The Sad End of Blockbuster," *International Business Times*, December 5, 2013, accessed February 24, 2017, http://www.ibtimes.com/sad-end-blockbuster-video-onetime-5-billion-company-being-liquidated-competition-1496962.

Figure 4.3. The New Disrupters: Platform Models with High Valuations that Provide Few Jobs

Company & Year Founded	Valuation	Revenue	Employees
Facebook (2004)	$328B	$5.2	13,000
Uber (2009)	$62B	$2B	6,700
Airbnb (2008)	$30B	$1B	2,368
Alibaba (1999)	$192B	$485B	36,450
Linkedin (2002, acquired by Microsoft in 2016)	$26B	$3B	9,900
Netflix	$46B	$7B	3,700

Source: Gerald Davis, *The Vanishing American Corporation* (San Francisco: Barrett-Koehler, 2016).

By far the best-known and most controversial of these so-called platform companies is Uber, the ride-sharing company. Uber has only several thousand employees (interestingly, little is known about its preferred employment strategy), but it also has well over a million drivers worldwide that it chooses to treat as independent contractors rather than employees. Most of Uber's drivers work part time. In the United States, outmoded labor laws deny to independent contractors many of the entitlements and legal protections—such as coverage by minimum wage laws and the right to unionize—that employees enjoy as a matter of course. Uber's decision to treat its drivers as independent contractors is particularly controversial and has stirred up multiple protests and lawsuits that are wending their way through the courts and regulatory agencies, both in the United States and in other countries. As we know, the wheels of justice turn slowly, so this issue is likely to take a long time to reach the U.S. Supreme Court for final resolution. Until the matter is settled, Uber and a growing number of other platform companies may be persuaded to adopt high-road employment strategies in their dealings with employees, but they almost certainly will not choose to do so in their dealings with those they consider independent contractors.

The Uber approach is not the only business model at work in the platform economy. In the ride-service industry, a New York City start-up called Juno[32] is attempting to compete with Uber by providing full-time good-paying jobs to employees. In Madison, Wisconsin, Union Cab[33] is trying to do so as well using a modified cooperative business form. Upwork is a platform business that links a variety of independent contractors with households that need various services and seeks to support a more open and balanced negotiated marketplace for providers and households seeking a variety of repair, maintenance, and technical services.[34] Care.com, the largest provider of home health care services, has adopted a code of conduct and a set of employment standards first developed by employee advocates called the Domestic Workers Alliance.[35] Shiftgig, a platform that matches certified staff for temporary work assignments for events such as concerts, one-time sports matches, and so on, aspires to do so in ways that help gig workers increase and broaden their skills and advance their career prospects.[36]

Which approach will dominate if, as is expected, the platform economy grows in the future? We cannot answer that question at the moment, but it is already clear that the choice will not be left entirely to individual entrepreneurs. Various groups of Uber drivers are now attempting to form unions or other groups to engage the company to improve their incomes and working conditions. Others are urging government regulators to extend the coverage of some or all employment laws to independent contractors. We will look at the pros and cons of these organizing and regulatory options for responding to the rise of platform business models in Chapters 6 and 7. For now it is sufficient to again return to a key theme in this chapter: Those who create and lead platform businesses have choices about how they compete and their choices have predictable consequences for the people who provide the services they sell. Indeed, there may be low- and high-road equivalents even in the platform economy. The question is whether the entrepreneurs who use these new business

[32] Polina Marinova, "Meet the Secretive Startup Trying to Steal Uber's Top Drivers," *Fortune*, March 28, 2016, accessed February 24, 2017, http://fortune.com/2016/03/28/juno-ridesharing-uber/.

[33] http://www.unioncab.com/.

[34] https://www.upwork.com/.

[35] https://www.care.com/.

[36] https://www.shiftgig.com/.

forms want to think creatively about how to address the needs of the multiple stakeholders who depend on and contribute to their success or wait for the countervailing pressures from service providers and government regulators to impose more of a balanced multi-stakeholder perspective on them.

Making It Happen: A Business Leaders' Action Plan

What concrete steps can business leaders take on their own to adopt the high-road approach in their own firms and to help spread it more widely across the country (and around the world)? Research and the experiences of the firms highlighted in this chapter have suggested the following list of actions.

- Examine your fundamental beliefs about the purpose of the firm and about the motives and aspirations of your employees. Are these beliefs consistent with building the kind of future you hope to see for your children or grandchildren? If not, perhaps it is time for a change not only in your beliefs but also the kind of employment strategies you choose to instill in your firm and would like to see in other firms.

- Learn about high-road strategies. The issues we raised in this chapter have been thoroughly researched over many years. If you lack the time to read the studies, have someone in your company do summaries. Consult with your human resource executives and managers who should be familiar with the research and with high-road firms in your industry and related industries with which you could benchmark. Touch base with some of the firms mentioned in this chapter to learn what is working for them and how they overcame the pitfalls they experiences on the way to the high road. If you're an entrepreneur beginning a start-up, put as much effort into learning about how best to manage your most important resource—your people—as you do into learning how best to handle your financing and your finances. Various organizations exist to help entrepreneurs get started on the high road. Find them and remember the advice to "pick your investors carefully."

- Be hard-nosed and analytical. Make changes for good business reasons; otherwise, they won't stick. Take a lesson from Jamie Dimon, the CEO of JP Morgan Chase: "While businesses such as ours are understandably cautious when it comes to expenses, there are good expenses (investments that will pay off in the long run) and bad expenses (waste and inefficiencies). We have never hesitated to invest aggressively [in our employees] if we thought it would improve our long-term prospects."

- Resist being a victim of financialization (especially by Wall Street analysts). Financialization can have a deleterious effect on the diffusion of the high-road approach. It fosters a short-termism that discourages investments that take a while to pay off. It also discourages the adoption of management practices that financiers see as being out of the mainstream. A firm with which we are familiar, for example, had grown to the point where it was considering an initial public offering. At the time, it had a well-functioning high-performance work system in place. Even after the core features and major advantages of the high-performance work system were patiently explained to their investment bankers, though, the bankers nonetheless advised the firm to scrap it. The bankers assumed that investors wouldn't understand it any better than they had and would downgrade the firm's initial stock price, perhaps by as much as 20 percent. The firm's executive team resisted this advice. They were firmly committed to and indeed quite proud of their high-road approach and were confident that it would prove its value over time, which it did.

- If necessary, push back against Wall Street analysts and potential investors by showing them why a high-road strategy will pay off for them in the long run. Show them the data on turnover, customer satisfaction, productivity, and employee ratings of your organization on public sites like Glassdoor and/or in "best places to work" competitions you might choose to enter.

- Step out. Experiment with disruptive models. Use independent contractors—the so-called Uber model—where it makes good business sense to do so, but be aware that there is a backlash against the lack of employment protections for independent contractors. So far we have no well-developed high-road models and only a limited number of examples to illustrate a high-road platform strategy that would work in this context. But sooner rather than later, platform businesses almost certainly will be called upon to deal with the backlash issue. Surely the first ones that figure out how to do it stand to gain a competitive advantage over those that are still mired in legal battles and controversy.

- When shifting from the low-road to the high-road approach, have a well-thought-out strategy for change. Be fully committed to the change. Be transparent; communicate your firm commitment to everyone in your company and to external stakeholders (e.g., investors) frequently and in many different ways to help overcome initial resistance. Consider starting with a pilot study in a small part or a single location of the company, if it is feasible to do so. Treat this as an experiment, being careful to track what is being learned and the results obtained. Be patient. Remember the "worse before better" dynamic and stay the course.

- Engage your employees, union representatives (if a union is present), and customers as allies and partners in building and sustaining your high-road strategies and your company's reputation. Remember the Market Basket example and do the mental calculation to determine what losing the support of these different stakeholders could do to the value of your business.

- Bring others along on the high-road journey with you. You'll need to engage with others in your industry plus a bevy of key stakeholders who can be of considerable help if they are on board.

    - Work with other leaders in your domain to encourage them to join you on the high road. Call out low-road firms, especially those that are cutting corners or violating employment standards and workers' rights.

- Engage with the leaders of relevant labor unions or other social activist groups. Treat them as allies and partners and not as adversaries. In the end, you share a common goal: improving the lot of current and future generations of workers.

- Partner with relevant educational institutions. The high-road approach works best when employees have well-developed, state-of-the-art technical and social skills. Encourage the development of a strong pipeline of talent with these skills. Be an active partner—share information about new technologies that are on the drawing board and, in turn, learn from the research educators have done and their knowledge of the skills these technologies will require and how best to develop them.

- Hold government leaders accountable, but at the same time work with them to develop a set of labor and employment laws and standards that are consistent with twenty-first-century realities and that support the diffusion of the high-road approach.

Video Resources

The following videos covering topics in this chapter can be found at http://iwer.mit.edu/speak-up-for-work/:

The Purpose of the American Corporation
The Changing Role of the American Corporation: Interview with John Reed
High-Performance Work Systems
The Market Basket Case with Zeynep Ton
Holding Global Corporations Accountable: From Nike to Apple
Alternatives to the Traditional Corporations
Entrepreneurship and Good Jobs: Interviews with Barbara Dyer and Scott Stern
Role of the Corporation in Different Societies
Interview with Nazma Akter, Labor Leader in Bangladesh

5

Married for Life: Workers and Educators

There is no doubt that the forces of globalization, financialization, and digitization will be with us for a long time to come and that the major players in the economy will continue to look for ways to capitalize on the opportunities and cope with the challenges these forces will bring. By now it is clear that we favor the view that these external forces are far from deterministic in terms of outcomes. Rather, as members of the work force, we, as individuals and as members of groups, can influence how events play out. The future workers experience will depend in large part on the actions they collectively take to shape a new social contract that suits their needs. In subsequent chapters, this theme of collective action is developed in discussions of the role of labor unions, other worker advocacy groups, and government policy-makers. But to a large extent the future workers experience also will depend on the actions they take as individuals to prepare themselves for whatever unfolds. This is the issue we address in the first half of this chapter. The world of work will continue to change, probably at an accelerated pace. Companies will come and go and the survivors will experiment with new variations of (we hope) high-road employment systems. New technologies will come along and lay waste to all or part of many existing jobs as firms learn to use new technologies (again we hope) to augment jobs and to create new and even better ones. In all of these changes, workers will need to keep pace. In a rapidly changing economy, the smart money is on those who are prepared. Specifically, the winners are most likely to be those who on their own initiative keep a watchful eye on emerging developments and do what is necessary to acquire and update essential knowledge and skills on an ongoing basis.

But ultimately it takes more than individual actions to sustain a labor force with cutting-edge competencies. Even the most alert and driven workers will need considerable support and assistance along the way. And it is fair to say that the existing ecosystem of education offerings on which they must depend is in a state of transition. The second part of this chapter begins by examining where things now stand in this respect. The main focus, though, is on using what has been learned so far to suggest a number of much-needed reforms that can and should be undertaken to improve the performance of the countless educational endeavors that include not only preschool programs, primary and secondary schools, and colleges and universities but also programs offered by companies, unions and other worker advocacy groups, and government agencies on their own and in various combinations.

The goal, of course, is to create a mutually beneficial collaboration between workers who clearly will need a first-rate basic education and then constant reskilling throughout their working lives and the myriad institutions that exist (or should exist) to provide the training these workers will require where and when they need it. It is also crucial that this training be presented in formats that are high quality, easily accessible, and affordable. The next-generation work force needs to be well prepared for a high-road, technologically augmented economy. And companies are likely to be more willing to create such an economy if they are convinced that the country's educational infrastructure is doing all it can to produce a work force that is constantly on the cutting edge—ready, willing, and able to keep the new economy advancing. Thus, both parties can do their part to build a new social contract at work.

The Work Force

As previous chapters have suggested, it is folly at this point to predict precisely how many of what types of jobs will exist in the years ahead. What is clear, however, is that whatever work needs to be done will be undertaken within a vast ecosystem of increasingly complex work arrangements. Not only will there be the usual range of employers—for-profit corporations, alternative forms of businesses (e.g., benefit corporations, small mom-and-pop stores) and other organizations (e.g., government agencies and nongovernmental organizations), and start-ups that choose to have standard employer-employee arrangements, there will also be an increasing number of start-ups and platform companies that have relatively few employees and much larger contingents of contractors (e.g., Uber) and a smaller but growing contingent of independents—contractors and gig workers—with no institutional affiliations. Intensifying competitive forces and technological developments are expected to bring considerable changes for the ecosystem's various components. Many of the current players will adapt and survive, but many won't. Further, many of the survivors will be struggling to develop and market an endless array of new and better products and services while also making continuous improvements in their internal operations. The experimentation with business models, organizational structures, technology-driven improvements, and workplace designs will be ongoing and intense.[1] For both organizations and individuals, the performance bar will continue to rise.

This is the nature of the ecosystem that today's work force already knows and tomorrow's work force will experience even more intensively. Obviously it has a number of implications for the workers involved, particularly with regard to their careers, competencies, and commitment.

Careers

During and even well after the post–World War II period, careers were largely defined in terms of onward and upward. Businesses and other bureaucracies (government agencies, the military, etc.) were organized horizontally by hierarchical levels and vertically by functions and it was expected that new entrants would settle into these structures and steadily progress to higher levels within their specialties and, if they had the right stuff, on to general management or executive positions. This system certainly was more prevalent among managers and professionals than among administrative or blue-collar workers, but even some members of the latter groups could expect to move into supervisory or perhaps lower-level management slots over time. In the 1980s and 90s, this traditional notion of careers began to break down as organizations flattened their hierarchies, cut back on training and development for managers, and eliminated large numbers of middle-management positions. Vestiges remain, although not many. Instead, careers increasingly are unfolding across many organizations and many experiences within organizations (less often up a ladder and more often up, across, and even down a so-called lattice of developmental opportunities).[2] And in a world with an increasing number of freelancers and gig workers, many careers largely unfold totally outside organizations of any kind.

In the traditional model, companies exercised considerable control over an individual's career, specifically in terms of when and where the employee would move. Serious career planning might have been done in schools (particularly for students bound for college), but from then on such planning became a fairly routine exercise.

[1] Christopher Worley, Thomas Williams, and Edward Lawler III, *The Agility Factor: Building Adaptable Organizations for Superior Performance* (San Francisco: Jossey Bass, 2014).

[2] Cathleen Benko and Molly Anderson, "The Lattice That Has Replaced the Corporate Ladder," *Forbes*, March 16, 2011, accessed February 17, 2017, http://www.forbes.com/2011/03/16/corporate-lattice-ladder-leadership-managing-hierarchy.html.

Currently individuals are much more on their own and career planning is a much more important, albeit challenging, process with at least two important components.

One component involves periodic assessments of career goals. Many tools exist to assist with this task. As we noted in Chapter 1, in our course we use a simple survey that asks students to rank their aspirations and priorities regarding work. What is it they really want from their careers? Do they want to make a lot of money? Do they want to work on important and challenging issues? Do they want a lot of autonomy? Do they want to help others? Do they want to be their own boss? Do they want to achieve a workable balance between work and home life? (Perhaps it is a sign of the times that we left off job security.) The information in Figure 1.8 and the student comments that accompany it reinforce an important point. There is a clear congruence between the high-road approach to workplace practices and the three fundamental career goals this group ranked number one most frequently: having a lot of autonomy on the job, working on important and challenging issues, and finding a balance between work and life. Further, while career goals can and do change over time[3]—making a lot of money, for example, generally increases in importance in the mid-career years when there are mortgages to pay and college tuitions in the offing—it is encouraging to note that the three fundamental career goals tend to persist in importance over time.

Clarity about career goals serves as a critical input to the second component of career planning: the exploration of career opportunities that not only hold promise for meeting these goals but also are consistent with a person's interests and aptitudes. Many tools are available for finding such jobs. In the course, we encourage students to use the U.S. Department of Labor's online tool, the O*NET Interest Profiler,[4] to plot their personal development plans. The Interest Profiler is particularly helpful because it provides a short (and free) questionnaire that assesses occupational aptitudes and uses that data to generate a list of occupations that align with those aptitudes. It also provides information about the educational requirements for different levels of jobs in various occupations. In addition, we strongly suggest that students discuss their goals and interests with people who are working in occupations they might be interested in

[3] Jeff Roberts, "Survey: 'Most Important' Career Goals Change with Age," July 31, 2014, accessed February 17, 2017, http://www.rasmussen.edu/student-life/blogs/main/most-important-career-goals-change-with-age/.

[4] https://www.onetcenter.org/IP.html.

as a sort of reality check. It is important that students' expectations square with the way things actually work on the ground. Career goals and interests may change over time, and occupations and jobs surely will change over time. So it will be necessary to recalibrate career development plans from time to time as misalignments occur.

This two-pronged approach to career planning relates directly to the issue of competencies.

Competencies

It's good for students and workers to gravitate toward occupations and jobs that are consistent with their career goals and interests. But it's also important, of course, for them to acquire the competencies these jobs require. Workers need a firm understanding of what these competencies are, an equally firm understanding on where they stand relative to these requirements, and a clear notion of the gaps between those two sets of data. Those gaps are what translate into their personal development needs.

While it's impossible to delineate the full range of competencies here, career guidance experts suggest that they tend to fall in three major categories.[5] One is technical skills, or the knowledge and capabilities that are specific to occupations or jobs. Increasingly, particularly in companies that gravitate toward the augmentation approach to automation, these competencies will relate to what generally are described as information technology skills such as graphic design, coding, simulation modeling, and data analysis. A second set of competencies relates to learning and innovation—or behavioral—skills such as creativity, critical thinking, teamwork, problem solving, communication, collaboration, negotiating skills, and the like. These are particularly important for those who find themselves in or wanting to join companies that choose to implement high-road workplace practices. In leading-edge companies where augmentation meets the high road, the emphasis is on so-called blended or hybrid jobs—for example, software engineers who do much of their work in teams using social media tools to collaborate and persuade. These tasks require high levels of both technical and behavioral skills. The third set of competencies has to do with so-called life and career skills—socio-emotional attributes such as initiative and self-direction, flexibility and adaptability, personal responsibility, accountability,

[5] Partnership for 21st Century Learning, "Framework for 21st Century Learning," March 2011, http://www.p21.org/storage/documents/1.__p21_framework_2-pager.pdf.

and agility—that help workers navigate in complex and rapidly changing work environments.

To this standard set of three categories, we would add a fourth set of ten competencies identified in research conducted at the Institute for the Future.[6] These competencies are elaborated on in one of the videos listed at the end of this chapter, so we simply note a few that add to those enumerated above. Certainly one of the most important is the capacity to function effectively in a globalized world (what the Institute of the Future calls "cross-cultural competency"), or the ability to work well with those from other cultures and function effectively in countries other than one's own. Closely related to cross-cultural competency is virtual collaboration, the capacity to work effectively with diverse teammates, often from around the world, who may never meet in person but interact only through collaborative software and social media contacts. And finally we would emphasize cognitive load management, or the ability to filter information for importance and to use technology in ways that simplify rather than overwhelm.

Note that most of these competencies relate to individuals who are or will be working in organizations either as employees or as contractors on site. However, these competencies relate in a very special way to those who are or will be freelancers and gig workers. In this context, we can do no better than to cite, as we do in Box 5.1, the experience of Steven Hill, who was laid off from his job at a Washington think tank and unexpectedly found himself working as a freelance journalist. He found the fact that he was a skilled researcher and writer with solid technical and behavioral skills was no longer enough. Instead, he found himself calling on a broader set of life and career skills—initiative and self-direction, flexibility and adaptability, personal responsibility, accountability, and agility—in a new and compelling way.

[6] Anna Davies, Devin Fidler, and Marina Gorbis, *Future Work Skills 2020* ([Phoenix, AZ]: Institute for the Future and the University of Phoenix, 2011), accessed February 17, 2017, http://www.iftf.org/uploads/media/SR-1382A_UPRI_future_work_skills_sm.pdf.

Box 5.1. Life and Career Skills of a Freelancer

Suddenly I was responsible for paying for my own health care [and] arranging for my own IRAs. . . . But that's not all. Not only was the pay per article or lecture not particularly lucrative, but I didn't get paid for those many hours in which I had to query the editors for the *next* article or lecture, or conduct research or interviews. It was as if I had become an assembly line worker who was paid on a per-piece rate; the "extraneous" part of my working day—rest and bathroom breaks, staff meetings, or time around the water cooler, usually paid time in a "good" job—had been stripped to the bone. Not to mention I no longer had paid vacations, sick days, or holidays, nor could I benefit from unemployment or injured workers compensation. Instead of receiving a paycheck from a single employer, now I had to track my many and varied sources of income, making sure unscrupulous editors didn't stiff me. In short, I had to juggle, juggle, juggle, while simultaneously running uphill.

Source: Steven Hill, *Raw Deal: How the "Uber Economy" and Runaway Capitalism Are Screwing American Workers* (New York: St. Martin's Press, 2015), 9, italics in original.

Hill pretty much had to go it alone. However, as we will see in the next section of this chapter (and in the next chapter), this should be less necessary in the years ahead.

Commitment

A while back, a writer by the name of Seth Godin penned a compelling headline: "In the face of change, the competent are helpless."[7] Competent people, he suggested, know how to do one thing really, really well. But what they don't know how to do—really, what they don't want to do—is change because change threatens their competence. This may work fine for a few niche players here and there, but for most of us in today's economy it is the kiss of death. What the world increasingly needs, Godin suggested, are serial incompetents, those who on their own initiative are willing to take the risks that come with constantly learning new things, those who are willing to live with the awful sense of vulnerability that comes with being on the upside of the learning curve not just once or twice but on an ongoing basis, those who are quick to master new tasks and restless enough to quickly move onto something new.

[7] Seth Godin, "In the Face of Change, the Competent Are Helpless," *Fast Company*, January–February 2000, 230–234.

Today we refer to serial incompetents with a more socially acceptable term, lifelong learners.[8] But the message is the same. And we hear the drumbeat all the time. New York Times columnist Thomas Friedman put it this way: "You need to work harder, regularly reinvent yourself, obtain at least some form of post-secondary education, make sure that you're engaged in lifelong learning, and play by the new rules while also reinventing some of them."[9] Or consider this quote from Randall Stephenson, the CEO of AT&T, which is in the throes of (yet another) major transformation: "There is a need to retool yourself, and you should not expect to stop." And here's what Ebony Bell, a product manager in AT&T's Atlanta office, has to say on the topic of the need for change: "I need to know what my competitors are doing. . . . I can't see myself staying with one product too long—it makes me feel like I'm not growing."[10] Bill Blase, the head of human resources at AT&T, puts it this way: "You have to do the 'what' and the 'how.' The 'how' is how you get along with people, [how] you achieve results with effectively partnering and teaming and leading change through [and with] others. . . . The 'what' is that you are not only proficient in your job but that you are reskilling to improve your capacity, continuing to learn, and that you are aspiring to go beyond where you are." John Donovan, AT&T's chief strategy officer, adds to the message: "You can be a lifelong employee if you are ready to be a lifelong learner. But you have to opt in. The company's responsibility is to provide the tools and platforms for employees to get there. . . . We need to make sure that anyone who leaves here [does not do so] because we did not provide them the platform—that it was their lack of motivation that did not make it happen."[11]

While we applaud AT&T's stance, it is clear that it is not enough to lay the responsibility for lifelong learning on individuals and companies alone. Ultimately, it will take more than these two parties, however motivated and committed they may be, to forge a work force capable of playing an active role in building the future social

[8] Marjan Laal and Payman Salamati, "Lifelong Learning: Why Do We Need It?," *Procedia-Social and Behavioral Sciences* 31 (2012): 399–403, http://www.sciencedirect.com/science/article/pii/S1877042811030023.

[9] Thomas Friedman, *Thank You for Being Late* (New York: Farrar, Strauss and Giroux, 2016), 205.

[10] These quotes are from Quentin Hardy, "Gearing Up for the Cloud, AT&T Tells Its Workers: Adapt, or Else," *New York Times,* February 13, 2016, accessed February 17, 2017, http://www.nytimes.com/2016/02/14/technology/gearing-up-for-the-cloud-att-tells-its-workers-adapt-or-else.html?_r=0.

[11] The quotes from Blasé and Donovan are from Friedman, *Thank You for Being Late,* 218.

contract at work. Workers enmeshed in a vast ecosystem of complex work arrangements that require lifelong learning need assistance and support from an equally extensive—and effective—ecosystem of educational opportunities.

Improving the Ecosystem of Educational Opportunities

The United States, as many other countries do, spends an enormous amount of money on various forms of education, both public and private. Currently these extend from the cradle to the grave—well, anyway, from preschool programs through primary and secondary schools and then on to colleges and universities and well beyond in the form of training and development efforts conducted by employers, unions and other worker-advocacy groups and various government agencies and independent providers. Sometimes these groups offer education on their own, but they also often do so in two- and three-way collaborations. The problem, however, is that these efforts largely operate on a piecemeal basis and often fail to deliver on their intended goals. The result is a highly differentiated work force. At one end of the spectrum are those whose competencies are, and in all likelihood will remain, totally in tune with the changing nature of technology and work. On the other end are those who at one stage or another will almost surely fall off the grid. If the country is to maintain a competitive economy well into the twenty-first century, this cannot continue. Instead, we need as many capable people as possible in the labor force over the long run. Just as is the case with high-road workplace strategies, the challenge is to learn from the educational endeavors that get the best results and then make every effort to assure that the lessons learned are diffused widely to where they can do the most good.

Formal Education

Let's begin at the beginning. Not long ago, one of us found himself sitting between two young mothers who were visiting the United States from China and Portugal. The conversation turned to how old their children were when they started going to school. The Chinese mother said that in her country all children go to state-supported programs as soon after birth as their parents will send them. The Portuguese mother said that where she lives, all children start government-supported pre-kindergarten programs at age three. Had an American mother or father been party to this conversation, it's hard to say what her or his story might have been. In the United States, access to high-quality preschool programs is unevenly distributed. For one thing, the programs are from universally available. Further, while high-income families generally are able and willing to pay for the programs that are available, low-income families usually cannot afford them without subsidies, and such subsidies are also far from universally available.

There is clear and convincing evidence that making the right investments in early childhood education pay significant economic and social dividends down the road.[12] Youngsters who start their primary school education behind the curve almost never catch up. This of course virtually assures uneven performance levels in later schooling and, in the long run, lower participation and success rates in higher education and other forms of lifelong learning and thus less access to and a lower likelihood of holding on to good jobs. This, in turn, exacerbates the already destructive level of income inequality in this country. Head Start, a federally-funded program, provides comprehensive early childhood education and other services to low-income children and their families. But because of funding and other limitations, this program reaches only about one million children a year. Thus, there is a need to expand the concept by creating a broader system of universally available and high-quality preschool programs and to make sure that resources are available to help all families gain access to these programs, irrespective of income. Without this foundation, it is unlikely that students and workers will ever be able to make full use of the remaining components of the ecosystem of educational opportunities.[13]

The problem is a bit different at the level of primary and secondary education. The United States invests heavily in public education. In the 2012–13 school year (the latest numbers we could find), the total was $620 billion, which breaks down to almost $13,000 per student. Only five countries spent more per student. The question, however, is this: What kind of returns are we getting? Overall, the results have been quite discouraging. High school dropout rates are declining, but they are still high—about 17 percent overall, and up to 50 percent in some schools. Further, far too many of those who graduate lack basic life and work skills. Consider, for example, a recent

[12] James J. Heckman, "Research Summary: The Lifecycle Benefits of an Influential Early Childhood Program," *Heckman: The Economics of Human Potential*, accessed February 24, 2017, https://heckmanequation.org/resource/research-summary-lifecycle-benefits-influential-early-childhood-program/.

[13] Josh Bivens, Emma Garcia, Elise Gould, Elaine Weiss, and Valerie Wilson, "It's Time for an Ambitious National Investment in America's Children," *Economic Policy Institute*, April 6, 2016, accessed February 17, 2017, http://www.epi.org/publication/its-time-for-an-ambitious-national-investment-in-americas-children/.

study by the Program for International Assessment of Adult Competencies. The study measured three essential work-related competencies—literacy, math, and technology—among adult populations in OECD countries. Overall, Americans' literacy skills were about average. Not surprisingly, respondents who went to college and graduate school did relatively well, while those with only a high school diploma or less performed at significantly lower levels than adults with the same level of education in other countries. In math, the results were even worse; Americans with a high school diploma performed at about the same level as high school dropouts in other countries. And the U.S. performance in the area of technology was abysmal. The American study participants finished in last place, just below Poland.[14]

A closer look tells a more nuanced story. There are many excellent public and private primary and secondary schools in the United States whose graduates consistently hold their own with the best in the world. Not surprisingly, these schools tend to be in relatively high-income communities that are able and willing to spend on education and where the students benefit from considerable parental support and assistance. In fact, much of the variance in student performance across schools disappears when the data are adjusted for demography and access to educational resources (e.g., parental education, the availability of books and technology at home, and whether English is the primary language spoken at home). Thus, while it is clear that there are problems, it isn't at all clear that they are entirely or even mostly related to the quality of education being provided. However, schools must educate all children irrespective of their backgrounds or preparation. There is constant pressure on underperforming schools to improve, along with considerable carrot-and-stick funding from federal and sometimes state governments and from foundations to experiment with innovative approaches (e.g., President Obama's "Race to the Top" initiative).

This is mostly a good-news story. There are many examples of reform efforts that have produced improvements, even in schools where most students come from low-income households.[15] One example is the Massachusetts Education Partnership

[14] Gabrielle Emanuel, "America's High School Graduates Look Like Other Countries' High School Dropouts," *nprEd*, March 10, 2016, accessed February 17, 2017, http://www.npr.org/sections/ed/2016/03/10/469831485/americas-high-school-graduates-look-like-other-countries-high-school-dropouts.

[15] "Bay Area School Reform Collaborative: Project Overview," *mrdc*, n.d., accessed February 17, 2017, http://www.mdrc.org/project/bay-area-school-reform-collaborative#overview.

(MEP), a consortium of state government officials, school boards, school superintendents, and teachers' unions that focuses on cooperative efforts to drive improvements in student achievement.[16] During its first two years, the consortium provided training and hands-on facilitators (neutral experts to help the parties use problem-solving techniques) to help teachers negotiate new labor agreements. These experts also provided intensive training for districts and local unions that were tackling specific reforms such as implementing new teacher performance evaluation systems and extending the school day/year. In Boston, for example, the consortium helped the parties reach an agreement to extend the school year by about a month, something they hadn't been able to do through traditional collective bargaining. It is assumed that efforts to improve teachers' performance levels, especially if students spend more time in the classroom, will result in better student performance, although the data on this are not yet in. So far, the MEP has worked with about one-third of Massachusetts school districts. The hope is that eventually it will serve as a model for the rest of the country to follow. A lot is riding on these reform efforts because at this point the graduates of even high-quality high schools can expect to earn little more than poverty wages if that is as far as they go in school (unless they happen to be Bill Gates or Mark Zuckerberg). To compete successfully for good jobs today and certainly tomorrow it is necessary to use the high school diploma as a ticket to further education, either at a community college for specialized technical training or at a four-year college and university or perhaps even on to graduate school.

Community colleges have long played an important role in work force education, and this must continue. Despite the effects of digitization on middle-skill jobs, there still is a need for commercial drivers, technicians, mechanics, electricians, plumbers, sales reps, and so forth, in part because these jobs will persist, although perhaps in modified forms, and in part because of the vacancies being created as baby boomers retire. The United States has an extensive network of vocational schools and community colleges to meet this need. Unfortunately, these institutions tend to be underfunded, they are often out of step with the competencies required in the twenty-

[16] Barry Bluestone, Thomas Kochan, and Nancy Peace, "Getting Along: A Better Approach to Public Sector Labor Relations—and Improving Schools," *Commonwealth: Politics, Ideas & Civic Life in Massachusetts,* April 14, 2015, accessed February 17, 2017, http://commonwealthmagazine.org/author/thomaskochan/.

first-century workplace, and they are underappreciated (when they can, students choose to attend four-year colleges and universities instead).[17]

While the challenges two-year schools face are many and complex, a great deal has been learned about steps that can be taken to improve their effectiveness, especially in terms of conducting training in workplace-specific skills (instead of, say, preparing students to transfer to four-year colleges or universities). Specifically, research and experience suggest that two-year colleges work best when:

- They work closely with employers in a region or industry (and often with unions or professional associations), not only to seek funding but also to ascertain precisely what types of knowledge and skills are required and then to jointly design and deliver programs specifically tailored to these learning objectives. It also helps to obtain up-front guarantees that all the students who successfully complete the custom-built programs will be offered good-paying jobs with participating companies.

- Classroom education is integrated with opportunities to apply new concepts and skills either on the job or, if this isn't feasible, in simulated work settings. Several studies have shown that this is the way those aiming for middle-skill jobs are likely to learn best.

- The training focuses on the specific knowledge and skills that are required for the jobs at hand but also goes further to work on the competencies required to prepare workers for longer-term careers with the employers or in the occupations involved.

One particular form of vocational training and certification that embodies all of these key features has a long and venerable history in the United States and other countries: Apprenticeships sponsored and funded by unions and/or by joint union-industry governing bodies. While apprenticeships have been most closely associated with craft occupations such as plumbers, carpenters, electricians and so on, some are

[17] Grace Chen, "7 Problems with Community Colleges and What Can Be Done about Them," *Community College Review,* May 20, 2012, accessed February 17, 2017, http://www.communitycollegereview.com/blog/7-problems-with-community-colleges-and-what-can-be-done-about-them.

also found in industrial trades such as tool and die making. Experimentation is under way in various places to create broader-based manufacturing apprenticeships that prepare workers to staff the evolving highly automated work systems that will likely dominate these industries in the future. Research on apprenticeships has shown they have achieved high rates of return in lifetime earnings for graduates and for the firms that hire them. One estimate is that on average an apprenticeship certification adds $250,000 to lifetime earnings. Other studies report rates of return of between 38 and 50 percent to employers that fund these programs.[18] Two well-known examples from the utility and the aerospace industries are summarized in Box 5.2.

Clearly, it is in the interest of everyone—business, the work force, unions, and the overall economy—to increase the number of apprenticeships open to young workers. Unfortunately, the United States has gone in exactly the opposite direction in recent years, in large part because the principle advocates and providers of apprenticeships—labor unions—have declined. From 1998 to 2010, the number of apprenticeship programs that were registered with the U.S. Department of Labor declined by 36 percent and the number of individuals enrolled in apprenticeship programs declined by 16 percent. As Peter Cappelli has pointed out, this decline is problematic since it means that currently less than 1 percent of the U.S. labor force receives apprenticeship training.[19] In contrast, nearly two-thirds of high school graduates enter apprenticeships in Germany, the country with the most advanced and widespread apprenticeship system.[20]

[18] Kevin Hollenbeck, "State Use of Workforce System New Impact Estimates and Rates of Return," paper presented at the Association of Public Policy and Management, Los Angeles, November 6, 2008, accessed February 17, 2017, http://research.upjohn.org/cgi/viewcontent.cgi?article=1006&context=confpapers; *The Benefits and Costs of Apprenticeships: The Business Perspective* ([Euclid, OH]: Case Western Reserve University and U.S. Department of Commerce, 2016), accessed February 17, 2017, http://www.esa.gov/sites/default/files/the-benefits-and-costs-of-apprenticeships-a-business-perspective.pdf.

[19] Peter Cappelli, *Will College Pay Off?* (New York: Public Affairs Program, 2015).

[20] Nika Lazaryan, Urvi Neelakantan, and David A. Price, "The Prevalence of Apprenticeships in Germany and the United States," Economic Brief 14-08, Federal Reserve Bank of Richmond, August 2014, https://www.richmondfed.org/~/media/richmondfedorg/publications/research/economic_brief/2014/pdf/eb_14-08.pdf.

While America is not likely to ever match the German scale, increasing the number and range of occupations covered by modern apprenticeship programs must be a top priority in building a new social contract. No other arena provides a better opportunity for joint business, labor, education, and government collaboration.

Box 5.2. Apprenticeship Programs in Utilities and Aerospace

The Center for Energy Workforce Development (CEWD) was formed in 2006 to deal with that industry's pending demographic cliff. CEWD is a nonprofit consortium of electric natural gas and nuclear utilities and their associations— the Edison Electric Institute, the American Gas Association, the Nuclear Energy Institute, and the International Brotherhood of Electrical Workers Union. Firms leading this effort, such as Duke Power, Pacific Gas and Energy, and Georgia Power report having hired thousands of line technicians, calculate positive rates of returns through reductions in hiring and contractor costs, and now have the needed replacements on hand as the predicted demographic cliff is upon them.

In 1988, Boeing and its machinists' union jointly studied other "best practice" joint training programs and then the company agreed to set aside $.14 per hour worked by the covered employees to build a joint training fund. That program blossomed over the years to feed Boeing's need for machinists, tool and die makers, computer technicians, and other growing jobs. More recently, Boeing and the machinists' union began working with the state of Washington, regional community colleges, and other local employers in its supply base to form the Aerospace Joint Apprenticeship Training Program. Together, these parties have developed common certification standards and curriculums for machinists, tool and die markers, and other workers who are critical to the future of the industry.

U.S. colleges and universities are arguably the most admired in the world, and there is no doubt that they are a linchpin in the country's ecosystem of educational opportunities. On balance, a college education is a fairly solid investment. According to a study by the Center on Education and the Workforce at Georgetown University, those with a bachelor's degree can expect to earn on average $2.8 million over a lifetime. Only 14 percent of high school graduates, 23 percent of college dropouts, and 28 percent of those with associate's degrees can expect to earn that much or more. In contrast, from 61 to 83 percent of those with some type of advanced degree can expect to match or exceed that number.[21] But there are no guarantees. A college education is an expensive proposition, and many students acquire considerable debt before graduation. Further, roughly half of the students who enter college never make it to graduation. They are left with the debt but without the degree. Clearly there is a need to find ways to reduce the costs of a college education and to help more of those who get into the system graduate. (It is also important to steer students with other interests and aptitudes elsewhere—e.g., toward technical training). Several organizations conduct surveys and provide online tools that allow comparisons across colleges and universities on a number of dimensions. These clearly show that there is wide variation in the performance of different colleges when that performance is measured in terms of graduation rates, job placement, or postgraduate earnings.[22] Of particular concern is the recent increase in the number of for-profit colleges. These schools are more expensive than public universities and have higher dropout rates. If ever the phrase "buyer beware" is warranted, it fits for students and families choosing which college to invest in and attend.

Obviously, there is need for reform at the college and university level. Thomas Friedman describes a fairly radical case of how one college embraced reform. Olin College is a small school in Massachusetts that specializes in the training of "exemplary engineering innovators," as the college's president, Richard K. Miller, puts it.[23] Nearly every aspect of the college, including the bylaws, has an expiration date. For example, the curriculum is constantly evolving and is thoroughly reviewed at least every seven years, when it must be either reinstated or revised. The college has a highly flexible operating structure. There are no departments and the faculty members are not tenured. (They are employed on renewable contracts that are regularly reviewed with no guarantee of reappointment.) The instruction is designed

[21] Anthony Carnavale, Stephen Rose, and Ban Cheah, "The College Payoff: Education, Occupations, Lifetime Earnings," *Center on Education and the Workforce*, Georgetown University, August 5, 2011, accessed February 17, 2017, https://cew.georgetown.edu/wp-content/uploads/2014/11/collegepayoff-complete.pdf.

[22] "Our First-Ever College Rankings," *The Economist*, accessed February 24, 2017, http://www.economist.com/blogs/graphicdetail/2015/10/value-university; College Results Online, accessed February 24, 2017, http://collegeresults.org/; 2016–2017 PayScale College Salary Report, accessed February 24, 2017, http://www.payscale.com/college-salary-report.

[23] Friedman, *Thank You for Being* Late, 222–224.

to prepare students to address the world's biggest technical, social, economic, and political challenges. Courses are deliberately integrative. A biology course, for example, is interwoven with one on the history of pandemics. And every student must complete a year-long, team-based, innovation-focused project with a corporate sponsor that provides both guidance and financial support. Friedman describes Olin's approach as "a real revolution in education [that soon] will be coming to a community near you."

Well, maybe. As long-term academics, we are well aware that revolution does not come easily to academia. However, there is no doubt that the experimentation should continue, certainly with a heavy focus on online instruction that leads to the attainment of credits and degrees rather than just certificates of completion. Surely we are not far away from the time when standard introductory courses for credit that are taught by celebrity professors from prestigious universities will be universally available via free or low-cost online courses.[24] Currently, the University of Wisconsin, to name but one, offers associate's, bachelor's, and master's degree programs that combine online instruction with competency-based testing.

Through the upheaval that technological change is sure to bring, though, most colleges and universities will cling to their basic mission: to prepare students to be literate and informed citizens and to ensure that the work force of the future has the basic knowledge and skills needed to enjoy successful careers in a globalized, knowledge-driven economy. This means that despite what some companies seem to think, institutions of higher learning are not in the business of producing just the right number of graduates with just the right types of skills precisely when they're needed.[25] Neither is it their job to develop competencies that will last a lifetime (evidence cited in Chapter 8 shows that even just a few years out of college, many graduates are employed in positions that do not make full use of what they studied.) It is imperative that we all keep right on learning.

Further Learning

As noted earlier, the system of formal education in this country is a model of efficiency and effectiveness compared with the picture that emerges with respect to what comes after a person graduates. At the risk of imposing more order than is

[24] Martin Ford, *Rise of the Robots* (New York: Basic Books, 2015), 143.

[25] Peter Cappelli, *Why Good People Can't Get Jobs* (Philadelphia: Wharton Digital Press, 2012), 52.

warranted, we propose that much of what currently occurs in the realm of further learning pertains to one of four situations: traditional careers, organizational restructurings, economic restructurings, and precarious careers.

As noted earlier, relatively few workers land or will land in traditional careers that involve employment with a single company or at most a couple of companies in mission-critical positions with planned paths up well-defined career ladders. Those who are on these career paths, however, are the people companies want to hold on to and continuously develop. So most medium to large organizations take charge of the process by providing various combinations of company-sponsored development activities—orientation programs, mentorships and coaching, on-the-job rotational and stretch assignments, in-house training programs, online courses, and on-campus executive development programs at well-known universities. MIT's Global Operations Program provides an example.[26] Back in the 1980s, MIT reached out to a number of major firms—Boeing, United Technologies, General Motors, Intel, and Dell, among others—to develop an industry-funded joint engineering and management curriculum tailored to product development, engineering, and human resources practices at these firms. Companies nominate employees to enroll in the program, although other students who do not work at member companies are admitted as well. The participating firms expect their employees to return to resume their careers. They also have first dibs on the unattached students. But there are no guarantees, so they must compete for them, not only with each other but also with other potential employers the graduates may choose to consider. So far, the program has produced over 1,000 graduates, and more than half of the unattached students have started solid careers with the participating companies. The idea, of course, is not only to help workers keep competencies current but also to help them develop the new competencies that future assignments will require. Most large organizations (corporations, government agencies, the military) have been engaged in these types of educational endeavors for a long time. They have continuously tweaked them over the years and consequently are very good at them.

But many of those who begin traditional careers don't stay in them. Organizations and circumstances change, and in the process well-planned career paths change as well. (We recall the perhaps-apocryphal tale of the young fast-tracker

[26] Thomas Kochan, David Finegold, and Paul Osterman, "Who Can Fix the 'Middle Skills' Gap?," *Harvard Business Review*, December 2013, accessed February 2017, https://hbr.org/2012/12/who-can-fix-the-middle-skills-gap.

who was being groomed for a particular position in a corporation, only to find a few years into his career that not only had the position disappeared but so had the unit it was in and even the country where it was located.) Also, organizations and workers change their minds. Either may decide that the traditional career approach was a big mistake and opt for an alternative path. Further, chances are most of the work force will never experience traditional careers even for a little while, either because they aren't an option or because alternative directions beckon. These workers experience many other forms of further learning that, alas, are considerably less systemic than the regimen available to those on traditional career paths.

One approach occurs when organizations restructure for whatever reason. When this happens, workers often are laid off and thrust back into the labor market and are forced to find other jobs, if they can. But even when employees aren't laid off, the changes certainly have significant educational implications. Consider again the case of AT&T we mentioned briefly above.[27] In 2007, Steve Jobs chose AT&T to be the exclusive network provider for the (then) new iPhone. This set the firm on the path of a major transformation from a phone company focused on hardware to a business solutions provider focused on software, big data, and analytics in head-to-head competition with the likes of Google, Amazon, IBM, and Accenture. Along the way, it became clear that a large chunk of the company's work force had skill sets that were rapidly becoming obsolete and that this situation would likely be ongoing. It was decided that rather than going the hiring and firing route, AT&T would take the high road by helping its employees reskill and hoping to retain as many workers as possible. The company decided to commit up to $250 million a year to the effort. It created a number of online and classroom training opportunities on its own and partnered with several universities—Oklahoma, Georgia Tech, Notre Dame, Stanford, and others—and with online learning venues such as Udacity and Coursera to develop additional specialized programs to meet ever-changing skill needs. In addition, the company worked with Georgia Tech and Udacity to develop an online master's degree in computer science and it negotiated a tuition of just $6,000 (as compared with a cost of $45,000 for the university's two-year on-campus program). In sum, employees at AT&T now have access to further learning opportunities (and to the likelihood of

[27] Friedman, *Thank You for Being Late*, 214–218. See also Quentin Hardy, "Gearing Up for the Cloud, AT&T Tells Its Workers: Adapt, or Else," *New York Times*, February 13, 2016, https://www.nytimes.com/2016/02/14/technology/gearing-up-for-the-cloud-att-tells-its-workers-adapt-or-else.html.

continuing employment) as long as they are prepared to engage in the necessary training, primarily on their own time, and are willing to embrace the concept of lifelong learning. AT&T sees this bargain as a potential model for the many other large companies that must constantly reinvent themselves to keep pace with more nimble competitors.

Box 5.3. The National Fund for Workforce Solutions

The National Fund is a dynamic national partnership of communities, employers, workers, and philanthropists that serves 30 communities. It connects low-wage workers to the education, training, and credentials they need to secure access to family-supporting careers. Since 2007, the National Fund for Workforce Collaboratives has served 42,299 workers and brought together more than 4,060 employers to establish 151 workforce partnerships across the country.

Source: National Fund for Workforce Solutions, http://www.nfwsolutions.org/.

Unfortunately, as we have seen, there are times when companies restructure and lay off large numbers of employees or completely shut down facilities, leaving their employees in the lurch. These are particularly challenging situations, especially in smaller communities where few alternative employment opportunities exist. While both the federal government and many state and local governments provide job search assistance and some funds for training laid-off workers, a particularly effective approach involves highly focused community efforts. Consider, for example, the consortium called Workforce Central in Wisconsin Rapids, Wisconsin. It is one of thirty-one National Fund for Workforce Solutions sites supported by the Incourage Community Foundation (see Box 5.3). Its effort, not atypically, grew out of an economic crisis, in this case brought on by the departure of the paper industry from the region. The consortium involves several local employers, Mid-State Technical College, and other community-based organizations. A key participant is the region's largest employer, the Ocean Spray cranberry cooperative. At a recent conference, Kirk Willard, the manager of Ocean Spray's local plant, and Kelly Ryan, president of the Incourage Community Foundation, described how the consortium works. A key component is the Manufacturing Partnership, which includes a group of local CEOs who meet regularly to share best practices regarding worker development and assessments to help businesses identify local skills needs. The consortium also offers grants and collaborative training to help workers develop the skills required to meet these needs. The collaborative training efforts include a variety of programs that Mid-State Technical College conducts that focus on developing manufacturing supervisors and certified production technicians and helping workers develop specific skills in areas such as safety, quality control, and maintenance. Overall, Workforce Central has been a great success.[28] For example, the newly trained workers have helped the local Ocean Spray plant grow and improve its productivity and quality to the point where it has won recognition as the company's best performing facility five years in a row. Other examples of equally successful regional training consortia are shown in Box 5.4. These and other programs are powerful examples of what can be accomplished when organized community efforts and resources are mobilized in the interest of promoting further education of the work force, and more of these efforts are desperately needed.[29]

[28] For the specifics, see Center on Wisconsin Strategy and Incourage Community Foundation, *Workforce Central Evaluation: January 2012* (Madison: Center on Wisconsin Strategy, University of Wisconsin-Madison, 2012), accessed February 24, 2017, https://incouragecf.org/wp-content/uploads/2012/01/WFC-Evaluation-January-2012.pdf.

[29] Kochan, Finegold, and Osterman, "Who Can Fix the 'Middle Skills' Gap?"; Pete Carlson, Sam Leiken, Stephen Michon, and Beth Siegel, "Raising the Bar on Skill Development: Findings from the Front Line," Regional Prosperity Project, June 2013, accessed February 17, 2017, http://futureworks-web.com/pubs/Raising%20the%20Bar%20on%20Skill%20Development%20Final.pdf.

Box 5.4. Examples of Regional Training Consortia

The Bay Area Workforce Funding Collaborative, founded in 2004, unites local governments, community colleges, and businesses in the San Francisco area. It has trained more than 700 unskilled and displaced workers for well-paid jobs with defined career ladders in the biotech and health care sectors.

Boston-based SkillWorks, founded in 2001, has placed more than half of the 500 displaced workers it has trained in new jobs. It has also upgraded the skills of more than 1,000 incumbent workers in health care, hospitality, property services, automotive services, and green industries.

The Wisconsin Regional Training Partnership (WRTP) includes unions, companies, and educational institutions. The partnership provides 40- to 160-hour training programs in technical and general skills for manufacturing, construction, and health care jobs. An experiment found that people who took part in the WRTP or other similar sector-based training had better outcomes than those in a control group of comparable people who were not in the WRTP: The WRTP participants were more likely to find steady jobs, work more hours, and earn a higher hourly wage. Indeed, they earned 29 percent more, on average, than the controls did in the year after the training.

Source: Thomas Kochan, David Finegold, and Paul Osterman, "Who Can Fix the 'Middle Skills' Gap?," *Harvard Business Review* 90 (December 2012).

Finally, we turn to further learning issues pertaining to the growing army those engaged in precarious careers—the so-called precariat[30]—in the United States and elsewhere. Precariat is a general term that refers to a social class of people whose economic well-being is shaky because of the lack of job security and the unpredictability of intermittent work and earnings. This group consists of three subgroups. One includes employees of organizations who are on other than traditional career paths and, therefore, are subject to layoffs or terminations and who often experience (sometimes extended) periods of unemployment as they move from job to job over the course of their working lives. The second subgroup is made up of those who are not employed by organizations but rather work as freelancers or gig workers. And the third is comprised of those who move between the other two subgroups; that is, in and out of employment at companies with intermittent periods

[30] Guy Standing, *The Precariat: The New Dangerous Class* (London: Bloomsbury, 2014).

of freelancing or gig working. What these three groups have in common is that once they leave the confines of formal education, they essentially are in charge of managing their own careers and personal development. As they move from job to job or gig to gig, they repeatedly will encounter potential employers and clients who will be interested in hiring or engaging them, but only if their competencies are on the cutting edge. Regrettably, the infrastructure available to assist these workers meeting these challenges is basically a hit-or-miss proposition; it is emerging but is still a maze to work through.

The most obvious source of further learning for this group occurs while they are employed or engaged on a company–based project. These situations provide excellent opportunities for networking with and cross-learning from peers within the same or related occupations. These periods also may provide some opportunities to take advantage of more formal training programs the company the freelancer is working for offers. But employers are reluctant to provide this type of training because of a classic market failure. Their thinking is as follows: "If I invest in training these employees or freelancers, my competitors won't have any incentive to do it; they instead will just hire away (some say 'poach') the talent I have developed. The result is that I will bear the costs while they reap the benefits." The well-known solution to this type of market failure is for employers in a given product or labor market to collaborate on essential training, thereby sharing the costs of the investment and then competing for the services of those who are trained. Examples of this solution are most often found in local labor markets and typically involve relatively low-skilled employees. An organization called the Local Employer Resource Network (LERN), for instance, provides a model and support for these types of community-based efforts.[31]

Alternatively, a few employers have set aside concerns about market failures and on their own have dedicated resources to developing employees who are not on traditional career paths. One of these is Starbucks, the well-known coffee company and coffeehouse chain.[32] More than 70 percent of its employees are students. The company realizes that most of them are not in for the long haul, yet statistically only half of them will ever complete their degrees without considerable help. Starbucks partnered with Arizona State University (ASU) to form the College Achievement Plan, which is available to all U.S.-based employees of Starbucks and its affiliated businesses

[31] http://www.thearc.org/thearcatwork-local-employer-resource-network.

[32] http://globalassets.starbucks.com/assets/39415f5a386a47259479e9f553246eef.pdf.

who are eligible for benefits. Prospective participants must apply to and be accepted at ASU. Once accepted, they immediately receive scholarships to cover part of their tuition and assistance with finding funding to cover the remaining costs. Participants choose from more than 40 undergraduate programs offered by ASU online. Each time they successfully complete 21 credit hours, the company automatically reimburses the full cost of the tuition as well as any mandatory fees that have been paid, provided they stay with the company and the plan. Thus far, over 6,500 employees have participated in the plan and 400 have gotten their degrees. Many other companies offer tuition assistance plans—some, such as the J. M. Smucker Company (purveyors of peanut butter, jelly, and other food products), pay 100 percent of the costs. Thus, another way for employees to move from more to less precarious economic circumstances is to go to work for Starbucks, Smucker, or one of the other forward-thinking employers that provide their workers with generous education benefits.

Perhaps the most precarious workers in terms of further learning are freelancers and gig workers, whose attachments to employers are often too tangential and short term to warrant offering much, if any, training. These workers face unique challenges. First, many of them find it necessary to keep current in terms of two sets of competencies, those specific to their occupational specialty and those pertaining to being what some call "the CEO of me" (that is, mastering the entrepreneurial skills associated with running what in essence are their own businesses). The second challenge is finding the time and energy to engage in further learning while constantly searching for new work even while performing on current assignments. Perhaps the best way to capture the labyrinth of approaches these workers use is to let them speak for themselves. Recently, Upwork put out a blog post that asked freelancers to respond to the following query: "How [do you] keep your skills up to date in a fast-paced, online world?" Following is a summary of the responses:[33]

- Several mentioned that they used online placement platforms such as Upwork, LinkedIn, CareerBuilder, and Monster.com, not only to look for new assignments but also to stay abreast of the skills that companies are looking for.

- Many also indicated that they relied on fairly traditional approaches such as seeking out "stretch" assignments that really challenge their

[33] https://www.upwork.com/blog/2016/08/keep-skills-up-to-date-online-world/.

skills. (As one freelancer put it, "Just put yourself in difficult positions and find ways to escape from them; that's the best way to learn new things."). Others noted that they belong to professional associations and, as hard as it can be, take time to attend conferences and seminars they offer. And, of course, several said they spend time reading current books and articles written by experts in their fields.

- Not surprisingly, the most frequently mentioned sources of learning involved online resources. These ranged from networking via the Freelancers Union's hives and Google Meetups to taking advantage of videos and tutorials that are available via YouTube and TED talks. Some indicated a preference for using blogs and Webinars because they offer the opportunity for two-way communication.

Perhaps the most interesting development in this respect is the emergence of university-based lifelong learning platforms that are relatively inexpensive or even free (although the Upwork blog respondents rarely mentioned these). For example, Udacity offers several non-degree programs on building websites, developing apps, and the like, while Johns Hopkins and Rice Universities have partnered with Coursera to offer an online course leading to a certificate in data sciences.[34] MIT, Harvard, and the University of California Berkeley are involved with EdX, which creates free online courses. Some states, worried about the shortage of the skilled workers they need to entice new businesses to their regions, are beginning to offer inducements for college dropouts to return and finish a degree. Tennessee, for example, is advertising its program heavily while providing grants and other sources of financial aid to keep the costs at an absolute minimum.[35]

Considerable experimentation is occurring in the realm of further learning, although to date little in the way of systemic assistance has emerged for those engaged in other than traditional careers. Thus, there is considerable room for action on the part of both the affected workers and those who care—or should care—about the future state of the U.S. work force.

[34] Friedman, *Thank You for Being Late*, 219–221.

[35] Melissa Korn, "Adults Lured Back to College," *Wall Street Journal*, December 27, 2016, A3.

Making It Happen: An Action Plan for the Work Force and the Ecosystem of Educational Opportunities

What follows are some recommended action steps for the next generation of workers and for key members of the educational infrastructure, broadly defined.

Current and Future Members of the Work Force

- Stay focused on emerging developments in the economy, your region, your industry, your occupation, and your employer (if you have one). Things will change; try not to be blindsided.
- Engage in ongoing career planning. Periodically reassess your career goals in light of emerging developments in the economy, your region, and so forth, as well as your changing interests. Explore new opportunities as they arise, assessing them in the context of your incipient interests, your aptitudes, and your capabilities.
- Be brutally frank with yourself about the state of your competencies. Spend time realistically assessing the skills that not only your current job or gigs require but also new opportunities you would like to pursue require. Also take a hard look at where you stand vis-à-vis those skill requirements. As you uncover gaps, convert them to concrete plans for skills development that include specifics about how you will find the time and resources to carry out these plans.
- Embrace the concept of lifelong learning. Understand that whatever formal education you were fortunate enough to acquire or are fortunate enough to be acquiring at the moment is only the beginning, the foundation for a life of proactive engagement and ongoing competency development.

Key Members of the Education Infrastructure

- Recognize that in today's dynamic world it simply is not enough to assume that workers and educational institutions alone can or should bear the burden for ensuring that the country has a globally competitive work force. To get the job done right will take a great deal more active involvement and cooperation among companies, labor unions and other worker advocacy groups, public policy makers, and independent providers of learning opportunities than has emerged thus far.

- As you partner with these groups, insist on the development of a universal preschool program that is both high quality and easily accessible and affordable by all, irrespective of income or geographic location. Strive to ensure that those who are charged with this foundational educational endeavor work under high-road conditions that include a living wage.

- Ditto with respect to insisting on more adequate funding for community colleges, especially for programs that are aimed at developing competencies for the twenty-first-century work force. Learn from accumulated experience to ensure that all programs that are aimed at relevant skills—both short-term and long-term programs—involve experiential as well as classroom instruction and offer reasonable assurances that jobs will be available for those who complete the programs.

- Rebuild and expand apprenticeship systems and adapt them to changing technologies and modes of production. Working together on training and skills certification might illustrate to labor, business, education, and government leaders the power of doing so on other topics.

- Continue to support efforts aimed at reducing the cost of a college education and increasing the completion rates of college students. Continue to accept the fact that colleges and universities have dual missions: to develop informed and literate citizens and to provide a foundation that will enable graduates to enter gainful employment and engage in lifelong learning. Encourage some institutions of higher learning to extend their mission by reaching out to nontraditional constituencies and by finding ways of systematically delivering vocationally relevant coursework at affordable prices to workers at all stages of their careers.

- Finally, all the talk about lifelong learning is well and good, but in reality it is just a hollow concept unless all members of the educational infrastructure fully embrace it. This means investing the time, effort, and money it will take to develop a truly integrated ecosystem of educational opportunities sufficient to make lifelong learning possible for the vast majority of the work force—people who either by circumstance or choice find themselves on other than traditional career paths. As we have shown, many initiatives currently exist, such as those that seek to address the needs of those who are in danger of losing their jobs (such as the AT&T's program of reeducation) or who are fortunate enough to live in well-organized communities (such as Wisconsin Rapids). More of this type of experimentation is clearly needed. The tough nut to crack, however, pertains to the growing ranks of freelancers and gig workers who at the moment are largely on their own. Upwork's survey suggests that thus far they seem to be relying largely on the efforts of high-tech entrepreneurs to meet their development needs. This is great as far as it goes, but it is unlikely that the challenge will be fully met without a concentrated effort by the full membership of the country's educational infrastructure.

Video Resources

The following videos covering topics in this chapter can be found at http://iwer.mit.edu/speak-up-for-work/:

Early Childhood Education for All
High School and College: Necessary but Not Sufficient
Life-Long Learning: Moving from Rhetoric to Reality
Careers and Competencies

6

How Labor Can Contribute to a New Social Contract

This chapter looks at what labor leaders and other worker advocates and representatives can do to contribute to building and sustaining a new social contract. The first question we address is why is labor (we use this term as a shorthand for both the formal labor movement as it exists today and the range of potential worker advocates and representatives who are experimenting with new forms of worker voice and representation) needs to be a full partner in shaping the future of work and a new social contract. As we noted in Chapter 1, we suggest that there are strong economic and political reasons why workers need a strong, independent, and forward-looking voice at the table. This has been true historically and perhaps is even more important now, given the long period of wage stagnation that paralleled the decline of union representation and the recent political and social tensions that have surfaced among those who feel they have lost their voice and are being left behind as the economy evolves.

But we recognize that this is a controversial issue. Not everyone shares our view that workers need a voice or need to rebuild the bargaining power they lost as unions declined. That is why we discuss the challenges involved in getting this point of view accepted by other key stakeholders in society. With this as a foundation, we then explore what existing unions and other worker advocates are doing or need to do to play a meaningful and constructive role in shaping the future of work and fostering a new social contract.

Do Workers Need a New and Stronger Voice? Lessons from Market Basket

Maybe the best way to preview and bring to life our more detailed answer to this question is to revisit what happened at Market Basket in the summer of 2014, when thousands of managers, clerks, truck drivers, warehouse workers, and customers revolted against a change in top management that threatened the future of a high-road business model, their jobs, and the prices and quality of service customers had come to expect.[1] Together this coalition of employees and customers, with broad support from the media and the public, did more to demonstrate how outmoded and inadequate America's labor laws and the standard ways for workers to have a voice are than any scholarly paper or government commission could ever do.

Labor laws are supposed to protect workers who speak out against unfair conditions and provide an avenue for collective bargaining. But these courageous employees had the good sense to ignore all the prevailing legal doctrines and constraints that rendered these laws essentially useless to them.

The National Labor Relations Act was passed in 1935 to end uncontrolled workplace conflict by promoting collective bargaining. In that era, there was a reasonably clear division of labor between workers and managers. The law specifies that if a majority of non-supervisory workers vote to form a union for the purposes of collective bargaining, the employer has an obligation to negotiate, but only over a specified set of issues: wages, hours, and working conditions. Business strategies that affect pricing, customer service, investments, and corporate governance are excluded from the purview of collective bargaining. They are seen as what came to be known as "managerial prerogatives."

But workers can get a only union if a *majority* of workers in a specific bargaining unit vote to form one. If only 49.9 percent vote to form a union, no one gets one and the employer keeps full control and has no obligation to deal with any form of collective employee voice. Let's take a look at how Market Basket employees challenged these practices.

First, all employees, from high-level executives to warehouse workers, joined together to protest, not for higher wages, but in support of their fired CEO and out of fear that the new CEO was abandoning the Market Basket business model of low prices achieved through high productivity, quality service, and good jobs. No separate bargaining units here and no separation of workers and managers—the managers led the walkout.

Second, these workers understood it was the decisions of the board of directors and owners that would change the business model to extract more cash from the company and that in turn that practice would likely lead to lower employee incomes and harsher working conditions. They went right to the heart of the matter and

[1] The summary that follows is taken fromThomas Kochan, "Market Basket and America's Outdated Labor Law," *Cognoscenti*, September 3, 2014, accessed February 17, 2017, http://www.wbur.org/cognoscenti/2014/09/03/market-basket-collective-bargaining-thomas-kochan.

completely ignored the traditional categories of collective bargaining. They made it their business to address issues that are considered "managerial prerogatives."

Third, these employees, especially the managers, took these actions at enormous risk. Supervisors and managers are excluded from the labor laws that protect workers and can be fired at will. In fact, seven of the managers who participated in these actions were summarily dismissed.

Fourth, the Market Basket employees did not request that management or the National Labor Relations Board recognize them as a bargaining unit before they embarked on their spontaneous collective action. The process of a government-run union election would have taken years, and if management resisted, trying to obtain one likely would have failed in the end. These employees used social media to organize their coalition without asking their employer for permission to do so or requesting government intervention.

Fifth, and perhaps most importantly, these employees had no ongoing channel within the firm for voicing their objections before the board took the actions that triggered the dispute. Their only recourse was a highly public confrontation that imposed severe costs on the business, disrupted and inconvenienced customers, and threatened loss of their jobs.

What can be done to update labor law to fit the modern workplace? Perhaps if the Market Basket employees had had a collaborative advisory body in place before the board made its harmful decision, one that had the right to information about the decisions that shape employment practices, the members of that body could have pointed out the likely consequence of taking that action and perhaps deterred the board of directors from making what turned out to be a disastrous mistake.

What the Market Basket case show us is that we have a lot to do to rebuild opportunities for people to express their voices and influence the conditions under which they work, support high-road business strategies, and serve their customers. Let's take a deeper look at the challenges and opportunities involved in doing so.

What Next for Labor?

The labor movement is dead. Long live the labor movement! We have heard both of these statements offered with equal conviction in recent years. Those who make the first argument point to the half-century of slow and then more rapid decline in union membership in the United States—and, to a lesser but still significant extent, in many other countries around the world. In 1970, unions represented 23 percent of the labor force; in 2016, that number was down to 10,6 percent of the overall labor force and to 6.4 percent of the private sector labor force. Clearly, changes in the economy, in technology, and in managerial practices outstripped labor's ability to adapt, innovate, and recapture the members it lost. Those who have given up on unions and many who actively oppose them expect this long-run trend to continue.

Those who chant the second statement fall into two groups. One blames labor's decline on the groups who want to destroy unions. These traditionalists simply want to bring labor back by fixing labor laws, mobilizing workers in opposition to labor's enemies, and restoring collective bargaining so it can once again become the engine of economic and social progress. The second group argues for more fundamental changes in workers' strategies. They don't want to walk away from collective bargaining; they want to redefine it to be a more flexible and less legally constrained force and to foster new ways of mobilizing workers and rebuilding bargaining power by taking advantage of advances in social media, digital technologies, and the knowledge, skills, and mobility of workers.

What's at stake here? Labor leaders like to tell us that labor brought us the weekend, and they are right. Moreover, as we noted in Chapter 2, the rise of collective bargaining after Congress passed the National Labor Relations Act was a major force in creating and sustaining the postwar social contract. The economy is now experiencing the long-term effects of the breakdown of wage norms and outcomes that began in the 1980s and has continued to today. As we noted in Chapter 4, the best estimates are that the decline of unions and worker bargaining power since the 1980s account for between 20 and 30 percent of the rise in income inequality. If history is a guide, workers will need some form(s) of collective voice to ensure that their interests and priorities are reflected in any type of new social contract.

The anger and frustration that surfaced in the 2016 U.S. presidential elections suggests another reason why some renewal of worker voice is needed. The strongest support for so-called populist candidates—Senator Bernie Sanders in the Democratic Party and Donald Trump in the Republican Party—came from regions and families that were hardest hit by job loss and a decline in family incomes. The same voting patterns were evident in 2016 among those in Britain who voted to leave the European Union—the so-called Brexit vote. In both countries, many of those who were most disgruntled were workers and families that had lost unionized jobs and have not been able to find work that pays equivalent wages and benefits. Clearly, healing the deep economic and political divisions laid bare by these voting patterns involves rebuilding the voices and bargaining power of workers, job opportunities for workers who have been most affected by recent economic trends, and confidence in the economy.

There is a core democratic value at stake here as well. It is hard to find a thriving democratic society that does not have a free labor movement. Labor is often one of the first groups to be suppressed when totalitarian leaders rise to power. Europeans, remembering the history of fascism, Nazism, and communism, tend to be more accepting than many Americans of treating labor as a "social partner" with a legitimate seat at the table in dialogues about the future of work and social and economic policies. The growth of populist political movements and candidates in Europe and America may be waking up leaders and citizens to the current threats to democracy and by extension the future of capitalism and market economies.

In Chapters 1 and 4 we suggested that worker voice and bargaining power are also an essential components of a high-road economic strategy. They are critical to diffusing and sustaining employer commitment to high-road ways of competing and distributing the benefits of productivity and economic growth. We will develop this point in this chapter by suggesting that rebuilding voice and bargaining power needs to be done in modern ways and in ways that benefit all stakeholders in society. This will require more collaborative and flexible forms of voice and representation than those that characterized much of collective bargaining in the era of the old social contract. Some early models of what we have in mind were developed and tested in the 1980s, and some of these continue to function well and evolve today. Other new approaches are emerging now. We will build on the earlier models and examine the new ones in some detail in this chapter.

But It Won't Come Easily

Achieving a renewal of worker voice will not be easy. Workers in the United States have ambivalent views of unions. The results of national polls have fluctuated over the years, but recently the polling organization with the longest history of asking about this issue found that 58 percent of Americans approve of unions, up from 48 percent in 2009. (See the long-term trends displayed in Figure 6.1). Another poll by the Pew Research Survey found that young workers are more supportive of unions than any other demographic group. Fifty-five percent of those aged 18 to 29 years have a favorable view of unions, compared to the overall average of 48 percent.[2] While this is a significant differential, it is still far from a ringing endorsement of unions on the part of young workers, much less on the part of the overall population.

Figure 6.1. Gallup Polls on Approval/Disapproval of Labor Unions, 1936–2015

Source: "America Support for Unions Continues to Recover," http://www.gallup.com/poll/184622/americans-support-labor-unions-continues-recover.aspx.

Students in the online class also expressed mixed opinions about unions. Some recalled bad experiences with unions that did little for them, some recognized the contributions unions have made in the past and the need for the role they play in advocating for improved conditions for low-wage workers today, and some saw the need for new ways for workers' voices to be heard. Box 6.1 presents a sample of what class participants had to say.

[2] "Mixed Views of Impact of Long Term Union Decline," *Pew Research Center*, April 21, 2015, accessed February 17, 2017, http://www.people-press.org/2015/04/27/mixed-views-of-impact-of-long-term-decline-in-union-membership/.

Box 6.1. Samples of MITx Online Students' Views of Unions

1. My brother-in-law had a union job at [a unionized company in the 70s and 80s] and he lived high on the hog, did very little work and generally enjoyed the good life while the rest of his family actually worked for a living. The pendulum had to swing back because unions took advantage of their collective bargaining power and were seen as being the demise of many good enterprising companies.

2. I believe low-wage [workers] are still struggling and unable to get their voices heard without fear of losing out. I think one reason for this is the low number of jobs that still have a union standing by them and protecting them.

3. I reckon that unions have to adapt to the changing work environment. I believe that in the past they focused too much on negotiations to get the best conditions to workers, but now I think that it's collaboration what is needed, how unions could spread the company's strategies and goals, so that everyone works to reach the same objectives, and in return how benefits would be spread across workers.

4. I think that there needs to be a balance between unions and companies. Obviously unions do a lot of positive work towards protecting the rights of workers and making sure they are not discriminated against by their employers. At the same time, companies do need to protect themselves and not be bullied by unions looking for more and more benefits that might hurt the shareholders of the company. I believe there needs to be a mutual give and take between unions and companies that work towards benefiting the major needs of all parties involved.

Survey numbers like the ones summarized above can be quite misleading. Overall attitudes toward unions are not as important as what individual workers want or need based on their specific work experiences. Surveys dating back as far as the 1970s have shown there is a representation gap in the United States: A substantial number of unorganized workers would join a union if given the chance to do so. In

1976, a national survey estimated that 33 percent of unorganized workers would join a union if they could.[3] Since then, that proportion has steadily increased to the mid-40 to 50 percent range, depending on the poll.[4] Thus, a substantial number of workers would like to be represented but simply cannot get access to a union.

Two primary reasons account for this paradoxical situation. First, U.S. managers have historically and by and large successfully opposed any efforts on the part of workers to organize a union. The most recent and most carefully constructed study on this topic found, for example, that only 20 percent of union organizing processes result in achieving a collective bargaining contract. If management strongly opposes the union, the probability of making it through the election and negotiation process to obtain a contract goes down to 10 percent. Thus, management prevails 9 out of 10 times when it opposes workers' efforts to organize.[5]

The results of the study summarized above illustrate the deeper problem: The labor law enacted in 1935 to protect workers' rights to unionize is badly broken and its representation election procedures no longer provide a viable channel for workers to gain access to collective bargaining. Indeed, the U.S. labor law places severe constraints on the form of worker voice that is protected. We will return to this issue in the chapter on what government might do to help build a new social contract.

One final obstacle holds back efforts to rebuild worker voice. A political impasse has persisted between labor and business over how to reform and/or update labor law for nearly forty years. In 1978, a modest labor law reform bill that the labor movement supported failed in the Senate by one vote. In 1994, a national commission was established to study and then propose changes in the law, but it failed to gain consensus or sufficient support to achieve change. In the first year of the Obama administration, the labor movement proposed changes through a bill called the Employee Free Choice Act. The bill passed the House of Representatives but died in the Senate without a strong champion.

3 Thomas Kochan, "How American Workers View Trade Unions," *Monthly Labor Review* 102 (April 1979): 23–33.

4 Richard B. Freeman. "Do Workers Still Want Unions? More than Ever," *Economic Policy Institute*, February 22, 2007, accessed February 17, 2017, http://www.sharedprosperity.org/bp182.html.

5 John Paul Ferguson," The Eyes of the Needle: A Sequential Model of Union Organizing Drives, 1999–2004," *Industrial and Labor Relations Review* 62 (October 2008): 3–21.

These legal reforms would not likely fare any differently in the near future than they have in the past. Something else will be needed if the political impasse over the legal rules governing worker voice and representation is to be broken.

What might some other approaches or new models of worker voice and representation look like? The answer at this point has to be a simple statement: It is too early to tell. But while we need to look for new approaches, it is also important not to throw the baby out with the bath water. So before turning to a discussion of the innovations emerging today, we need to note one from the past that proved its value and warrants consideration in the future. We are referring to partnerships between labor and management that aim to engage employees and management in ways that both improve company performance and support good jobs and improvements in employment conditions.

Lessons from Previous Innovations

The best way to illustrate the power of labor management partnerships may be to describe the boldest labor-management experiment in modern history: the Saturn Corporation.

The Rise and Fall of Saturn

By the mid-1980s, some labor leaders and their management counterparts in highly unionized firms could see that things had changed early in that decade. They began searching for new directions in labor-management relations. See Box 6.2 for a simplified comparison of the traditional labor management model of the 1940s through the 1970s and what some of us called a transformed model based on the innovations in labor-management relations that were occurring in the 1980s.

Figure 6.2. Traditional and Transformed Models of Labor Relations

Level of the Firm	Traditional	Transformed
Strategic management	No union role; all decision rights retained by management	Labor-management consultation; information sharing or board representation
Collective bargaining	Arm's-length contract negotiations	"Interest-based" negotiations (i.e., the use of problem-solving processes)
Workplace engagement	Grievance procedures available to appeal violations of contract terms	Employee engagement in processes to improve the workplace and company operations

Source: Adapted from Thomas A. Kochan, Harry C. Katz, and Robert B. McKersie, *The Transformation of American Industrial Relations* (Ithaca, NY: Cornell ILR Press, 1994).

Progressive labor leaders such as United Auto Workers (UAW) vice-president Don Ephlin and his corporate counterpart at GM, Al Warren, fostered a series of innovations that laid the foundation for the spread of new work systems and labor-management relations. Their two boldest experiments were negotiations to create a joint venture with Toyota called New United Motors Manufacturing, Inc. (NUMMI)[6] and the creation of a new autonomous division of GM called the Saturn Corporation.[7]

[6] The story of NUMMI is best told in John Krafcik, "Triumph of a Lean Production System," *Sloan Management Review* 30, no. 1 (1988): 41–52; Paul Adler, "The Learning Bureaucracy," in *Research in Organizational Behavior*, edited by Barry M. Staw and Larry L. Cummings (Greenwich, CT: JAI Press, 1992); and Wilhelm Wilms, Alan J. Hardcastle, and Deone M. Zell, "Cultural Transformation at NUMMI," *Sloan Management Review* 36, no. 1 (1994): 99–113. The July 17, 2015, episode of NPR's program *This American Life* also provided information about NUMMI's history; see http://www.thisamericanlife.org/radio-archives/episode/561/nummi-2015.

[7] The material on Saturn is taken from Saul Rubinstein and Thomas Kochan, *Learning from Saturn* (Ithaca, NY: Cornell ILR Press, 2001). If you prefer to see a video summarizing the history of Saturn instead of reading through this text, check out excerpts from *Whatever Happened to Saturn*, Merrimac

The goal was to make both NUMMI and Saturn learning laboratories for GM, the UAW, and other business and labor leaders.

The Saturn story starts with Irving Bluestone, Don Ephlin's predecessor and mentor in the UAW. Bluestone was among the first of America's labor leaders to embrace the idea that all workers should have a voice in improving quality and productivity and the way they do their jobs. He initiated what came to be called Quality of Work Life experiments in a number of General Motors plants in the 1970s and negotiated with his GM counterpart to insert language that included these experiments in the UAW-GM national collective bargaining agreement in 1979.

In 1983, Warren told Ephlin that GM could not produce small cars competitively using UAW workers in the United States and would have to outsource all future small car models to some lower-cost country. Don's response was to say "Let's see if we can do the job." He and his GM counterpart created a Committee of 99 composed of engineers, accountants, mechanics, assembly-line workers, and other specialists to study the world's best practices in work systems, labor relations, and corporate strategies and structures. The result was a decision to create Saturn as "a new kind of company and a new kind of car."

In the Committee of 99's proposal, Saturn would create a union-management partnership from the bottom to the top of the organization. Gone would be the detailed job classifications and complex work rules that dominated traditional plants. Instead, the Saturn labor agreement would be only twenty-eight pages long. UAW wages and benefits would be achieved but in a different way: workers' pay packets would be a combination of standard wages, production bonuses, and profit sharing. Workers would assemble cars in teams jointly led by UAW and management representatives and the UAW would have co-management responsibilities across all management functions, from manufacturing to sales to product development and from the shop floor to the CEO's office.

This radical departure from UAW and GM management traditions was hotly debated, both within these organizations and across the country. But its critics were silenced early on when Saturn was well received by its customers. It achieved the highest quality and customer satisfaction ratings of any American-made automobile.

Films, 2008. You can find it on our website at http://www.speakupforwork.com/the-next-generations-american-dream-and-how-to-achieve-it-lessons-from-our-mitx-course/.

Only Lexus and Infiniti, luxury vehicles that cost twice as much as a Saturn, received higher customer ratings.

But the resentment of traditionalists within both the UAW and GM led to Saturn's downfall. First, they cut back on the investment dollars that had been built into the original business plan for building the number of vehicles needed for Saturn to be sustainable and profitable for the long run. Then they debated for over three years about whether to approve funding for second-generation models as the initial ones approached the end of their product cycle. By this time, the president of the UAW was Steve Yokich, a union traditionalist who opposed the ideas of employee participation and collaboration with management that his predecessors Bluestone and Ephlin had championed. As a result, Saturn's sales began a long slide downward. (See Box 6.3 for vignettes that further illustrate why Saturn was doomed.) By the time Saturn's product line was finally refreshed with a popular and high-quality SUV and a new but less-well-received sedan, the die was cast. Employee morale at Saturn had plummeted, as had productivity. Basic math ruled: If you maintain employment security when demand for your product declines, average cost per product goes up and profits turn to losses.

Internal leadership rivalries erupted and Saturn lingered on life support until GM gave up and closed it down as part of its restructuring during the government bailout process in 2008.

Box 6.2. The Leaders Who Doomed Saturn

Three vignettes from personal experience illustrate how GM and UAW leaders' lack of interest in or understanding of Saturn contributed to Saturn's downfall:

1. When Saul Rubinstein (co-author on our Saturn book) and Kochan finally got an "audience" with GM CEO Richard Wagoner to get his views on Saturn, we were astounded to learn how little he knew about how Saturn worked. We asked him, for example, "What do you think about having UAW representatives as co-leaders of the work teams at Saturn?" His answer was "Really, is that what they do there?" We were on the one hand amazed and on the other startled to learn how little he knew about the organization he was abandoning.

2. When we briefed several GM executives on the results of our work at Saturn in preparation for sharing the results with a larger group of GM managers, one executive said: "Can you just not tell them these results come from Saturn? They will close their

ears to anything that comes from there. The reality is that people at Ford have learned more from Saturn than we have."

3. In 1995, while sharing a ride to the airport from a meeting in Washington with UAW president Steve Yokich, Kochan got an earful about why Yokich would never let GM allocate another nickel to Saturn. Among his other comments was that "it's not the f____ kind of union I come from and not the kind we want associated with the UAW." The rest is history. The membership of the UAW was once 1.5 million; now it is down to 390,000.

Mike Bennett, the leader of the UAW local union in the early years of Saturn, discussed with our online students the lessons he took away from his experience. In response to a student's request for his advice about how to build the organizations and jobs of the future, he wrote:

> "Saturnize America" is my short answer to this broad and important question. Without profit there is no capability for "good" salaries, or "job security." . . . I would focus on hiring responsible people who want to grow, contribute and achieve individual and corporate success with customer focus. Exceed the customers' expectations! Develop a mission statement that every employee has input into, adopt a set of values that are lived up to, be fair and honest with everyone, treat everyone with respect and dignity, insist that everyone be valued-added to the mission and product. Invest in training people and expanding everyone's skill sets, [lead] by example, be a resource to the workers on the plant floor, practice active listening, . . . empower people with decision-making and hold them accountable.
>
> As for the future of jobs? In a global economy the rules have changed. Capital and technology have no borders and no national interests. . . . There is no job or income security without customers and there are no customers without customer focus! Always remember, in today['s] global economy, nothing is permanent.

Saturn and NUMMI were by no means the only innovators, and they influenced other union and management leaders to be similarly innovative. Across almost all industries, from steel to telecommunications to airlines, similar smaller-scale innovations in worker participation and labor-management collaboration played out in the 1980s and 1990s. Some of these innovations endure today. The most notable and the largest in scale, impact, and durability is an ambitious labor and management partnership between the health care giant Kaiser Permanente and the coalition of unions representing the majority of Kaiser employees (see Box 6.3).[8]

Box 6.3. The Kaiser Permanente Labor-Management Partnership

In 1997, the CEO of Kaiser Permanente (KP), the president of the AFL-CIO, and leaders of the coalition of the unions representing employees at KP created what was to become the largest, most long-standing, and most innovative labor-management partnership in the nation's history.

Over its first decade, the partnership helped turn around Kaiser Permanente's financial performance, built and sustained a record of labor peace, and demonstrated the value of using interest-based processes to negotiate national labor agreements and to resolve problems on a day-to-day basis. Among its most significant achievements included negotiation of a system-wide employment and income security agreement for dealing with workers affected by organizational restructurings. This agreement provided a framework that supported the introduction of electronic medical records technology on a scale that has made Kaiser Permanente a national leader in this area. In 2005 negotiations, the parties committed to bring partnership principles more fully to bear to support continuous improvement in health care delivery and performance by forming "unit based teams" (UBTs) of nurses, technicians, doctors, and service providers.

Since then the parties have achieved significant progress in integrating the partnership into the standard operating model for delivering health care by expanding UBTs throughout the organization and demonstrating that high-performing teams that engage employees contribute significantly to improving health care quality and service, reducing workplace injuries, improving attendance rates, and achieving high levels of employee satisfaction with KP as a place to work and a place to get health care. As a result, Kaiser Permanente is now one of the nation's leaders in the use of front line teams to improve health care delivery.

[8] Thomas Kochan, Adrienne Eaton, Robert McKersie, and Paul Adler, *Healing Together: The Labor-Management Partnership at Kaiser Permanente* (Ithaca, NY: Cornell University/ILR Press, 2009).

Source: Thomas Kochan, "The Kaiser Permanente Labor Management Partnership: 2009-2013," http://mitsloan.mit.edu/group/template/docs/iwer/FINAL-KPreport130947.pdf.

But these innovative examples did not spread across other industries and have had a hard time surviving. The sad reality is that labor leaders couldn't make up their minds on whether to support and champion the new approach or remain committed to the twentieth-century model that has led to their downfall. Those of us who participated in two top-level AFL-CIO "Future of Work" study groups and in the Commission on the Future of Labor Management Relations during the Clinton administration watched in dismay as advocates of new ideas, including Tom Donahue, secretary-treasurer of the AFL-CIO, and Lynn Williams, president of the United Steelworkers, were unable to convince enough of the traditionalist leaders to chart a new direction. Instead, the labor movement remained ambivalent—not opposed to collaborative methods if a leader wanted to pursue this option but not willing to get out in front and champion a new approach for all workers and their unions.

This suggests a key lesson for the future: Workers need labor leaders (and management leaders) that are unambiguous champions for innovation. Ambivalence or institutional inertia will lead to further union decline and the alienation of young workers and the public. Experiments with totally new models like the one that developed at Saturn and then was adapted to fit at Kaiser Permanente can demonstrate and test new approaches. When they work, they should become part of the standard labor portfolio. The key is to learn from new models and then devise ways to support and spread them across the work force.

Current Innovations

Now fast-forward to today. Labor union membership in the private sector has fallen to below 7 percent of the private sector labor force. Union leaders and other worker advocates have awakened to the reality that if they don't find ways to innovate, the labor movement as we have known it for the last century will become a footnote in history. This reality is stirring a significant amount of innovation inside the labor movement and among labor advocates who are acting on their own or in coalition with organized labor. Let's take a look at what they are doing.

Renewal Strategies from within the Labor Movement

We can start by looking at what existing unions are doing to rebuild. The AFL-CIO recognizes the need for change. In 2013, President Richard Trumka made a passionate call for innovation and a more inclusive labor movement (see Box 6.4).

Box 6.4. AFL-CIO President Richard Trumka on New Labor Strategies

It is time, my friends, to turn America right side up. And to turn America right side up, we need a real working class movement. And if that's going to happen, we—our institutions—have to do some things differently.

We must begin, here and now, today, the great work of reawakening a movement of working people—*all* working people, not just the people in this hall, not just the people we represent today—but everyone who works in this country, everyone who believes that people who work deserve to make enough to live and enjoy the good things in life.

We heard that all over America, workers are organizing in all kinds of ways, and they call their unity by all kinds of names—workers' unions, associations, centers, networks.

We heard that people want to be part of our movement but it's too hard to join—that we have to change so that our unions and our movement are open to everyone—to anyone who wants to join together for a better life. And today we're going to do that.

We heard that we have to change to reflect the times. The AFL and the CIO merged over 50 years ago, before the jumbo jet, before the cell phone, before the internet. We need to organize ourselves in ways that fit with the jobs people do now and how our economy works now.

And finally, we heard we have to make our unity real with action—we have to be able to organize on a large scale, in the workplace and in political life—quickly, efficiently, decisively. And with a strong, independent political voice.

And in everything we do, we have to join together with partners and allies who share our values and our vision for America. An America of shared prosperity. An America where you don't surrender your humanity, your dignity, your rights when you come to work. An America where we honor each individual, while understanding that connecting with each other, supporting each other—solidarity and community—are what give life meaning.

Throughout history, the energy and hopes of young workers have powered progress: If we are going to move forward, we must truly open our doors to the next generation.

If we are going to move forward, we must make our movement and our leadership as diverse as the workforce we speak for.

If we are going to move forward, we must move forward together—immigrants and the children of immigrants.

Source: "Remarks by AFL-CIO President Richard L. Trumka, 2013 AFL-CIO National Convention Keynote, Los Angeles, California," September 9, 2013, http://www.aflcio.org/Press-Room/Speeches/Remarks-by-AFL-CIO-President-Richard-L.Trumka-2013-AFL-CIO-National-Convention-Keynote-Los-Angeles-California.

Many individual unions are looking for new ways to build their membership and establish relationships with employers that can benefit both workers and firms. The United Steelworkers union, for example, promotes alternative corporate forms that involve employee ownership or worker cooperatives. In Portland, Oregon, and elsewhere, the Laborers International Union has played a key role in training workers in skills needed for green jobs, such as brownfield remediation and building weatherization. The Culinary Workers Union in Las Vegas has one of the best and most extensive training and outreach programs to help immigrants enter the service industries (gaming and others) that dominate that city and region. And the AFL-CIO has enrolled over three million nonunion workers through an affiliate called Working America[9] that does not engage in collective bargaining but provides information about and mobilizes members in support of local and national policy issues and political initiatives.

A number of unions are also targeting some of the large employers in low-wage sectors that heretofore have vigorously opposed the efforts of employees to organize. In 2012, a few hundred fast-food workers went on a day-long strike in New York City to protest low wages, irregular schedules, and poor working conditions. In 2015, this grew into organized one-day protests supported by the Service Employees International Union (SEIU) across 150 cities. In November 2016, low-wage workers in cities across the country again took up the call by protesting in support of the

[9] http://www.workingamerica.org/.

"Fight for $15."[10] Similarly, the United Food & Commercial Workers International Union, which organizes employees in grocery and retail stores and in the food-processing and meat-packing industries, created OUR Walmart, an organization of nonunionized Walmart employees who are working toward establishing higher wages, regular schedules, health care benefits, and respect on the job.

Outreach to Young Workers

The AFL-CIO has created a young workers' group that meets at local levels across the country. In 2015, over 1,500 of these young activists came together in a national conference. The group has developed its own economic policy strategy called Common Sense Economics.[11]

One of the biggest assets organized labor can offer young workers is access to its numerous apprenticeship training programs. As noted in Chapter 5, apprenticeships have among the highest rate of return in lifetime earnings of any training programs in the country and they get consistently high ratings from employers because they produce talented workers with the skills, work habits, and motivation employers need. Unfortunately, the number of apprenticeships has declined in the United States over the years as unions have declined. Today, however, unions are taking a number of initiatives to rebuild existing apprenticeship programs and build new ones in industries that are growing, such as those using advances in robotics, 3D printing, and other emerging manufacturing technologies. In addition, many unions are actively seeking to enroll more diverse young people in order to better reflect the makeup of the communities in which unions are active.

Training and Apprenticeship: Can Labor Be a Supplier of Choice?

One of the best examples of the type of expanded training and apprenticeship programs called for above is the Wisconsin Regional Training Partnership (WRTP) in Milwaukee, Wisconsin. Since its start in the 1990s, the WRTP has been

[10] Noam Scheiber, "Eyeing the Trump Voter, 'Fight for 15' Widens Its Focus," *New York Times*, November 29, 2016, accessed February 17, 2017, http://www.nytimes.com/2016/11/29/business/economy/fight-for-15-wages-protests.html?_r=0.

[11] James D. Gwartney, *Common Sense Economics: What Everyone Should Know about Wealth and Prosperity* (New York: St. Martin's Press, 2016).

a leader in providing training and career services in the Milwaukee region. It describes itself as industry led, worker centered, and community focused.[12] In 2015 it served 2,563 workers, trained 1,055, linked 631 to employment or careers, and placed workers in manufacturing jobs at a mean wage of $15.63 and in construction jobs at a mean wage of $22.83. An independent evaluation in 2003 found that its program trainees earned $4,500 more than those in a control group, worked longer hours, and had greater access to fringe benefits. It recently launched an industrial apprenticeship program with support from the U.S. Department of Labor that could be a model for apprenticeship initiatives in manufacturing.

Farm Labor Organizing

Historically, immigrant workers employed on a seasonal basis to do farm work have been very difficult to organize. The most famous successful case of farm worker organizing was in the 1960s, when Cezar Chavez built the United Farmworkers Workers (UFW) and secured labor agreements with grape and lettuce growers in California.[13] Since then two other notable efforts have used similar boycott and public campaigns to organize farmworkers. One is the Farm Workers Organizing Committee (FLOC), which Baldemar Velasquez founded in Ohio in 1967.[14] It took nearly a decade of concerted effort, including putting considerable pressure on the Campbell Soup Company, the purchaser of tomatoes from farmers in the area, to achieve an innovative tripartite agreement that covered farm workers, growers, and food processors. Part of the agreement created a neutral private commission chaired first by former Secretary of Labor John T. Dunlop and more recently by Wilma Liebman, the former chair of the

[12] http://www.wrtp.org/. See also "Manufacturing: Better Opportunity and a Stronger Economy," a joint report by the Wisconsin Regional Training Partnership and the University of Wisconsin Center for Wisconsin Strategy, 2013, accessed February 17, 2017, http://www.cows.org/manufacturing-better-opportunity-a-stronger-economy.

[13] Marshall Ganz, *Why David Sometimes Wins: Leadership, Organization, and Strategy in the California Farm Worker Movement* (New York: Oxford University Press, 2009).

[14] "About FLOC," *Farm Labor Organizing Committee, AFL-CIO*, accessed February 20, 2017, http://www.floc.com/wordpress/about-floc/. See also W. K. Barger and Ernesto M. Reza, *The Farm Labor Movement in the Midwest: Social Change and Adaptation among Migrant Farmworkers* (Austin: University of Texas Press, 1994).

National Labor Relations Board. The commission worked to resolve grievances and workers' complaints that were covered under the agreement and to mediate negotiations.

Another farm worker example is the Coalition of Immokalee Workers in Florida.[15] This is a human rights group that is not part of the formal labor movement but engages in a broad array of negotiations and advocacy programs that seek to upgrade and enforce labor standards for tomato growers in Florida and other states in the Southeast. Like FLOC, this coalition reached agreement with a group of retailers such as Taco Bell, Walmart, Whole Foods, and Subway to pay a small premium price (a penny a pound) that is passed on to the workers.

These examples from agricultural workers are interesting for several reasons. First, farm workers are not covered under the National Labor Relations Act; thus, each group had to either lobby for their own law at the state level (as the UFW did in California) or obtain agreements from growers and retailers without the benefit of a law. Second, they used combinations of public campaigns and economic pressure to bring attention to their low wages and harsh living conditions. They combined labor organizing with human rights campaigns that church groups, fair food coalitions, students, and anti-slavery advocates supported. Third, they all recognized that traditional bilateral negotiations with growers would not be sufficient to reach agreements; they needed to engage in multilateral negotiations with growers and key purchasers of their products in the supply chain.

These innovative steps are encouraging and need to be expanded and sustained. And they need unambiguous support from labor leaders. Remember the lesson of Saturn and other innovations that started and sputtered in the past: Ambiguity or, worse, ambivalence toward innovation will ensure that labor's decline will continue.

But there is a harder lesson to accept as well. Today, "labor," or more generally the work force, does not speak with one voice and will not be motivated or satisfied with one form of voice and representation at work. Instead, organized labor will need to work in coalition with multiple groups, largely grassroots in nature and perhaps more linked by occupation or social identity (race, gender,

[15] "About CIW," *Coalition of Immokalee Workers*, accessed February 17, 2017, http://www.ciw-online.org/about/.

ethnicity, etc.) than by occupation in order to identify the forms of voice and sources of bargaining power that serve the needs and interests of today's workers.

The Search for New Approaches to Worker Voice

Now let's take a look at some of the ideas and innovations that have been springing up either outside the established labor movement or in coalitions of community and labor leaders. As we will see, a variety of approaches are being tried that range from union-like organizations to community coalitions to entrepreneurial ventures that seek to learn from experiences of business start-ups to efforts to use digital technologies to develop apps that can be used to engage employers and/or to better match workers to jobs.

"Alt-Labor" Initiatives

The search for ways to supplement traditional union models that rely primarily on collective bargaining is sometimes referred to as "alt-labor."[16] One of the early innovators in this category is the Freelancers Union,[17] which Sara Horowitz formed in the 1990s after she found out that her employer was classifying her as an independent contractor who was not entitled to the same rights and benefits as her colleagues. Today, the Freelancers Union has over 250,000 members from a variety of occupations, including consultants, writers, nannies, and software developers. While it can't negotiate a collective bargaining agreement because it members are mostly independent contractors who are not covered under the National Labor Relations Act, the Freelancers Union functions like a union in many ways. For example, before the passage of the Affordable Care Act, it had its own health care plan and now it provides information on alternative plans that fit its members' needs. It also provides networking support and information exchanges about job opportunities and training programs and engages in political and policy activism in support of independent contractors.

[16] Josh Eidelson, "Alt-Labor," *The American Prospect*, January 29, 2013, accessed February 20, 2017, http://prospect.org/article/alt-labor; "Alt-Labor Rising: Organizing Walmart and the Fast Food Industry," *Demos*, November 18, 2013, accessed February 17, 2017, https://talkingunion.wordpress.com/tag/alt-labor/.

[17] https://www.freelancersunion.org/. See also Josh Eidelson, "Alt-Labor," *American Prospect*, January 29, 2013, accessed February 17, 2017, http://www.npr.org/2013/05/20/185559550/the-future-of-the-workers-movement. If we are not mistaken, Josh was the first to use the term "alt-labor."

The National Domestic Workers Alliance is another example of a new and growing organization. The alliance brings together organizations that represent over 10,000 nannies, home health care workers, and housekeepers. Its local affiliates have made significant inroads into improving standards in domestic work. In states such as New York and California, for instance, the alliance's efforts have brought about the passage of a Domestic Workers' Bill of Rights that provides for overtime pay, protection from discrimination, and paid leave days. These are things we usually think of as fairly typical or as normal rights on the job. As we noted briefly in Chapter 4, Care.com, which describes itself as the world' largest marketplace for linking families needing services to caregivers, has now adopted this Bill of Rights as part of its business and customer service model.[18]

Another visible and innovative union-like organization is the Restaurant Opportunities Centers United (ROC). ROC was created in New York City after the Windows on the World Restaurant in the World Trade Center was destroyed on September 11, 2001. Its model of training, raising wages, improving working conditions, and promoting sustainable food standards has expanded to numerous cities nationwide. Through these activities, ROC aims to change industry-wide standards. Importantly, it also conducts research on the industry that becomes incorporated into consumer education campaigns that seek to make customers aware of working conditions in the restaurants they frequent. This gives customers the opportunity to become allies for food-service employees who are seeking better wages and conditions in their industry.[19]

Perhaps the most ambitious new effort to invent the next-generation workers' organization is The Workers Lab, an "accelerator" for new worker advocacy organizations that is modeled after similar accelerators for business start-ups. As conceived by David Rolf, the president of the SEIU local in Seattle, The Workers Lab provides seed funding to initiatives that show promise for rebuilding worker bargaining power in new ways, have the potential to grow to a scale large enough to generate improvements in working conditions for significant numbers of people, and have financial models that eventually can be sustainable without outside subsidies. In its first year, The Workers Lab funded four

[18] "Company Overview," *Care.com*, accessed February 17, 2017, http://www.care.com/company-overview.

[19] Saru Jayaraman, *Behind the Kitchen Door* (Ithaca, NY: Cornell University/ILR Press, 2013).

projects.[20] One of these is the Restaurant Opportunities Centers. Another is coworker.org, an organization that helps individuals and groups organize campaigns to fight injustices at work through online petitions and other information-based strategies. A third is the Workers Defense Project, which educates immigrant workers and contractors about their rights and advocates for them about issues such as wages they are owed, improvements in safety and health conditions, and other basic labor standards. The fourth is WorkAmerica, an online recruiting network that links apprentices, vocational/technical schools, and employers in order to streamline the talent development and placement processes for those with critical skills in construction and health care. More recently, The Workers Lab has supported "hackathons" to develop training programs for creating worker advocacy start-ups and digital apps that workers can use to communicate about their working conditions and share ideas on a private server.

The rise of Uber, Lyft, and other platform-based taxi services have triggered several approaches to organizing drivers even, though drivers who are treated as independent contractors do not have the protections of labor law. Sherpashare[21] is a for-profit start-up that serves as a trusted intermediary between drivers, the ride-sharing companies, and their customers. It provides drivers with an online forum where they can track their expenses and income and compares overall driver incomes across cities. It also hosts forums where drivers can share information and can gather together in their cities. However, it does not seek to organize drivers into a collective force to negotiate with Uber or Lyft. Other drivers' groups such as the Drivers Network[22] and the Uber Drivers Forum[23] have sought to mobilize workers to engage with the companies. In 2016, Seattle's city council enacted legislation allowing drivers to join a union. The AFL-CIO issued a charter to the National Taxi Workers Alliance[24] that allows it to organize taxi drivers, and various unions are actively recruiting ride-service drivers. In 2016, the

[20] http://theworkerslab.com.

[21] https://www.sherpashare.com/.

[22] http://www.driversnetwork.org/.

[23] "Uber Drivers Forum," *UberPeople.net*, accessed February 17, 2017, http://uberpeople.net/.

[24] Robert Struckman, "Taxi! Taxi! Cab Drivers Form Unlikely Union," n.d., *AFL-CIO*, accessed February 17, 2017, http://www.aflcio.org/Features/Innovators/Taxi!-Taxi!-Cabbies-Form-Unlikely-Union.

International Association of Machinists and Aerospace Workers union created a "guild" to support drivers who work for Uber in New York City.[25] Uber has agreed to consult with this organization on a variety of practices and provides some financial support for its operations. Time will tell which, if any, of these different options proves to be sustainable and responsive to these drivers' needs.

Despite all of this activity, as David Rolf, one of the SEIU leaders of the Fight for $15 campaign has noted,[26] three central questions loom large. Most of these efforts are resource intensive, and there are questions about whether they can be replicated and expanded to reach large numbers of workers. Second, all of them have yet to figure out a business model that provides a sustainable source of revenue. Third, none have yet demonstrated that the sources of power they draw on, as innovative as some of them have been, are adequate substitutes for the power that collective bargaining generated in years past. Perhaps the next generation of worker entrepreneurs will build from these examples and discover and create ways to meet these challenges.

Worker-Centered Apps: Putting Digital Technologies to Work

As the coworkers.com example suggests, various groups are beginning to experiment with using digital technologies as new sources of power. Another example was started as OUR Walmart by the United Food and Commercial Workers Union. The founders of OUR Walmart now operate as an independent group outside the labor movement. In 2016, they released an app called WorkIt that enables Walmart employees to communicate with each other and get expert advice on how they might address workplace problems they experience. Workers can also query the app to get information on employment laws. The app uses IBM Watson artificial intelligence technology to provide some of the answers, while humans provide answers to more complex questions. Walmart executives have urged employees to not use the app.[27] Worker-centered apps like this one are

[25] https://drivingguild.org/.

[26] David Rolf, *The Fight for $15: The Right Wage for a Working America* (New York: New Press, 2016).

[27] Sarah Nassauer, "Wal-Mart Tells Workers: Don't Download Labor's Group Chat App," *Wall Street Journal*, November 15 2016, accessed February 17, 2017, http://www.wsj.com/articles/wal-mart-tells-workers-dont-download-labor-groups-chat-app-1479214510.

likely to grow in number and in the services they support and perhaps are early entries into a new, next-generation techno-mediated workers' movement.

Community-Based Initiatives

Perhaps the biggest organizing efforts outside the formal labor movement are the worker centers now found in about 240 localities that support immigrant and other low-wage workers by providing training in job skills and English and legal assistance for workers who are injured on the job or experience violations of other employment laws. In Chicago, New Orleans and many other cities, worker centers help workers pursue claims of wage theft, instances when workers believe they have not been paid the legal minimum wage or for hours worked that are due overtime pay.[28]

Another type of community-based organizing relates to local efforts to transform employment policies in specific localities where public funds are used in economic development projects or in contracting for city services such as trash removal or recycling. These campaigns often tie provisions for wage standards, training, and hiring policies to development projects that are funded through taxes or public subsidies. The overarching goal of many of these campaigns, similar to what we see with worker centers, is to transform jobs in particular sectors or a local economy into high-road jobs.

Other worker advocates have negotiated with elected officials and in some cases with representatives of the business community to increase minimum wages, establish living wage targets, or provide community benefit agreements (agreements negotiated with private developers to hire local residents and contribute to other community improvements). Currently, over 140 cities and counties have living wage laws that bring local wages above the federal minimum of $7.25 per hour and tailor them to the local cost of living. One of the largest community benefit agreements was formed in 2004 between a broad coalition of community and labor organizations and the Los Angeles International Airport, which was undergoing an $11 billion modernization project. The agreement provided for $15 million in training funds, addressed environmental and health concerns related to airport traffic, and created increased opportunities for

[28] Janice Fine, *Worker Centers: Organizing Communities at the Edge of the Dream* (Ithaca, NY: Cornell University/ILR Press, 2006); Kim Bobo, *Wage Theft in America* (New York: The New Press, 2011); Kim Bobo and Marien Cassillao Pabellon, *The Worker Center Handbook: A Practical Guide to Starting and Building a New Labor Movement* (Ithaca, NY: Cornell ILR Press, 2016).

minority- and women-owned businesses. The LAX Community Benefit Agreement included provisions for training and hiring local residents, thereby ensuring that the economic gains from development would be distributed throughout the community.

Exit Rather than Voice

Years ago, an influential book by economist Alfred Hirschman suggested that workers, consumers, or citizens can gain power by either (or both) exercising voice in their existing organizations or by leaving an employment relationship in favor of a better opportunity elsewhere.[29] The former was the preferred model for labor in the past and it still is in the present. But the rise of digital technologies is opening up new ways to explore whether exit is a new, albeit indirect, source of power workers can use to penalize employers who offer low-road jobs and reward those who offer better options.

Some of the organizations taking this approach would not call themselves labor organizations at all. Glassdoor[30] is perhaps the biggest example. It is a website where employees can rate their employers. Glassdoor has accumulated millions of ratings of the wages, employment conditions, views of CEOs, and other information on virtually all of the largest employers in the country. Indeed,[31] LinkedIn,[32] PayScale,[33] and other platforms are sources of similar information. More specialized labor market intermediaries or vendors that match labor supply and demand have emerged and are growing in number as well. There are, for example, several hundred "vendors" that place travel nurses in temporary jobs in hospitals around the country. And there are online forums available where travel nurses can rate the quality of job placements of both the vendors and the hospitals. Other matching services are active in labor markets for dental assistants and for home repair (e.g., Task Rabbit[34]) and a variety of other personal and business services.

[29] Alfred O. Hirschman, *Exit, Voice, and Loyalty* (Cambridge, MA: Harvard University Press, 1970).

[30] https://www.glassdoor.com/index.htm.

[31] http://www.indeed.com/.

[32] https://www.linkedin.com/.

[33] http://www.payscale.com/.

[34] https://www.taskrabbit.com/.

While each of these help expedite job matching and many provide ratings of employers or purchasers of services, none have yet taken up the task of mobilizing employees or contract workers to more directly influence employer practices. This may be the next step needed to transform exit options into a significant new source of worker power that can ratchet up employer practices and working conditions.

Making It Happen: An Action Plan for Labor

While a lot of innovation is under way, will it grow to a scale large enough to fill the void in worker voice and bargaining power? Will innovations find sustainable sources of revenue (i.e., business models) that serve the role that union dues played in funding and sustaining traditional unions? Or will those who oppose unions for ideological or other reasons continue to win the battles over worker organizing until they push existing unions and worker advocates to revolt in more forceful and large-scale ways? Time will tell. However, one of the lessons from our historical review in previous chapters is that we should never pronounce the labor movement dead. The industrial unions that emerged in the 1930s out of the ashes of the old craft unions spurred both new and existing unions to adapt to the changing industrial order of the day. A similar rebirth is possible, and if the examples cited in this chapter are an indication, it may be happening today.

But while we believe that independent, innovative, and powerful networks and organizations that mobilize and give workers a voice on issues are important to workers, they are also essential for achieving and sustaining a high-road economy and are critical to building a new social contract. The organizations that will help the next generation of workers gain a powerful voice need to build and learn from past union successes, but they cannot be built in the mirror image of the unions that were dominant in the mid- twentieth century and have declined in membership and influence.

Here then are the lessons we take away from both paths not taken in the past and from the actions labor leaders and advocates within and outside of the labor movement are now in the midst of taking once again in efforts to fill today's void in worker voice and bargaining power. Labor leaders, worker advocates, and workers themselves need to:

- Support labor-management partnerships such as the ones started at Saturn and carried forward at Kaiser Permanente. Both demonstrated their value: Their members have had a voice in driving improvements in productivity, quality, and customer service (i.e., supporting high-road business strategies) while achieving and maintaining industry-leading wages and benefits. A key lesson learned from these innovative models is that labor needs to champion and support these partnerships.

- Experiment with other types of partnership models such as the broadly representative (all employees at a worksite, in a company, or in some cases in an industry) works councils found in Europe. The UAW tried to put such a consultative body in place at the Volkswagen plant in Tennessee, but that attempt failed. More efforts along these lines are worth pursuing.

- "Trust but keep your powder dry" still applies. The campaigns of fast-food workers and others that seek to give voice and the right to representation to workers in low-wage jobs, immigrants, and others who are subject to mistreatment must be a central part of labor's mission and strategy. Partnerships with willing employers are sustainable only if there is an understanding that the alternative is labor's willingness and ability to draw on more traditional sources of power to protect and advance member interests. These more traditional pressure tactics will need to be used in dealing with employers who oppose the efforts of workers to gain representation or who choose to keep unions or professional associations at arm's length.

- Break out of the chains of existing labor law. Market Basket employees and customers showed the power of spontaneous collective actions that ignored what labor law or union organizing traditions had to offer. Indeed, the Restaurant Opportunities Centers explicitly state that they are not organizing a union for the purpose of collective bargaining as provided by labor law precisely because they do not want to be constrained by the law's "ossified" rules.

- Continue building coalitions with community-based organizations that are advocating for worker rights and better working conditions for low-wage workers, immigrants, and others who face multiple challenges at work and at home.

- Supplement workers' efforts to find new ways to improve voice and bargaining power in their current jobs through networking about good job opportunities with other employers/contractors. Make exit and mobility a source of power. This will require providing labor-market services such as information on good job opportunities, access to training, and so forth so that workers can maintain their union membership throughout their full career.

- Become the supplier of choice for high-skilled workers. Open existing apprenticeships to all workers and offer them lifetime services that provide incentives for graduates to retain their union membership as they move across jobs and employers—regardless of whether or not their employers have collective bargaining agreements. The online education arena is an untapped potential resource that labor can use to expand the educational services it provides to members. Labor could take a page from business's playbook by negotiating group discounts for online courses that generate certificates or credits toward advanced degrees. Making lifelong learning a key part of what unions offer might be one way for them to build a sustainable financial and membership growth model.

- Keep experimenting, innovating, and incubating new ideas. Keep inventing new ways to use digital technologies, including artificial intelligence, social media, and crowd-sourcing strategies to generate new approaches to worker voice and bargaining power. The next-generation labor movement is most likely to come from ideas, energy, and leadership from the next-generation workforce. Make room for them to lead the way.

Video Resources

The following videos covering topics in this chapter can be found at http://iwer.mit.edu/speak-up-for-work/:

Interview with AFL-CIO Secretary Treasurer Liz Shuler
The Rise and Fall of Saturn
The Decline of Unions in the United States
Emerging Models of Worker Advocacy
Emerging Forms of Work/Organizations
Uber Case Study
Shiftgig: An Interview with Vivienne Ming

7

How Government Leaders Can Contribute to a New Social Contract

Now we come to perhaps the most controversial question of the day: What should government do to help build and support a new social contract? We address this question as the Trump administration is taking office. The question on many people's minds is: Where will they take national labor and employment policies? Will they help heal the deep wounds laid bare in the election? Will they get the economy growing rapidly and create enough new good-quality jobs to offset those lost to globalization, technological change, declines in worker bargaining power, and so forth? Or will policy turn toward deregulating the labor market, weakening labor standards, attacking unions, and ignoring opportunities to encourage more firms to support high-road strategies?

Our approach in this chapter is to not wait and see. Instead, we lay out a strategy for the future that we believe would support movement toward a new social contract that gets the economy working better for all the key stakeholders—the work force, business, labor, our communities, and our society. We will leave it up to current and future leaders and, indeed, to all who have a stake in and share responsibility for building a better future to decide if they want to pursue this approach. We hope they will.

Government's Primary Responsibilities

The primary responsibilities of government leaders with regard to work are to manage macroeconomic policies to promote job growth and employment continuity, enact and enforce employment policies that balance the needs and interests of workers and employers in ways that align with the nation's values and public interests, and promote and support innovations in employment practices that increase productivity and the number of good jobs.

We have learned that most national policy innovations are first incubated and tested at the local and state levels or in the private sector. Building on these local examples would be a good starting point. But nothing short of a major transformation of current labor and employment policies, leadership, and enforcement will be adequate. What follows is a blueprint for federal action that is based on the lessons from the New Deal and on more recent local innovations. It is informed by and supports the efforts of business, labor, education, and the work force that are already showing pathways toward a new social contract.

Macroeconomic Policies: Investing for the Future

We have to start with macroeconomic policy. Here we need to take to heart the central macroeconomic policy lesson of the 1930s. It took the massive investment of the government in gearing up production for World War II to end the Great Depression. Now, in the twenty-first century, it will take an infusion of sustained investment—perhaps a mix of public and private dollars—to meet the nation's needs for more high-quality jobs.

One of the best investments government could make would be to begin rebuilding America's deteriorating infrastructure. The American Society of Civil Engineers estimates that America has a $3 trillion backlog of repairs to the nation's infrastructure. Conservative economists such as Martin Feldstein, liberal economists such as Paul Krugman, and many in between endorse this strategy for creating jobs and energizing the economy.

The reason there is such broad-based support for infrastructure investments is that they generate positive long-term economic returns and create good jobs. Berkeley economist Laura Tyson estimates that an investment of $100 billion would generate approximately two million jobs.[1] Moreover, many of these jobs would be good ones that require a middle-skill apprenticeship or the equivalent technical training and/or professional engineering or management degrees.

Most of the proposals for an infrastructure initiative call for a mixture of private and public capital as the initial source of funds. If Congress is unwilling to act on its own, the private sector could take the lead in raising at least some of the necessary capital. As early as 2011, the labor movement announced that it was prepared to commit up to $10 billion in pension funds to an infrastructure initiative. Given their special interest in fostering a competitive and growing economy, Wall Street firms could be called on to build a substantially larger pool of funds. Then business and labor could jointly propose that government further leverage these funds in ways that would lower the interest rate costs associated with infrastructure projects.

Having business and labor as co-investors has another potential benefit. Their joint oversight could help ensure that the organizations doing the work adopt state-of-the-art employment practices and create high-quality jobs and that the projects are managed efficiently and completed on time, within budget, and safely. This would

[1] Laura D. Tyson, "A Better Stimulus Plan for the U.S.," *Harvard Business Review* 89, nos. 1–2 (2011): 53.

both produce the sustained economic stimulus that is needed and demonstrate how business and labor can work together in a sustained way on a project of significant national interest. They have done so in the past during world wars and in the space race that met President Kennedy's goal of putting a man on the moon.

Infrastructure is only one potentially attractive investment opportunity. Equally important for building an innovation economy would be restoring investments in science and engineering to the levels of previous decades. MIT president Rafael Reif has been a consistent and articulate advocate for government investment in scientific research.

> Innovation is fueled by a long-time partnership between the federal government and the nation's scientists and engineers. Since World War II, federal funding for science has led to important technological breakthroughs and contributed mightily to our national defense. Over the long term, as much as three-quarters of economic growth may be attributable to innovation and technological change. . . . [But] federal R&D as a percentage of GDP—essentially, our societal commitment to research—has fallen from 1.3 percent in 1979 to 0.8 percent in 2013. Our competitors are going in the other direction.[2]

What about manufacturing? The loss of about one-third of the nation's manufacturing jobs has moved many families out of the middle class, devastated communities, and fueled a good deal of the anger that dominated the discourse and results of the 2016 elections. Some of these jobs were lost to automation and some to globalization. Can jobs lost to globalization be brought back to the United States? Can further declines in well-paying manufacturing jobs be avoided?

The good news is that some firms report that they are "reshoring" manufacturing work as the full costs of producing and transporting their goods to market tip the scales back in favor of domestic manufacturing. One firm that provides a software tool to support these calculations estimates that between 2010 and 2016, 265,000 jobs were either reshored or kept in the United States after employers had considered whether to send them abroad.[3] It is hard to know if this is an accurate estimate given the private nature of these business decisions and the mix of considerations that influence them. For example, a number of the cases cited in this estimate note that purchasing firms such as Walmart are now providing incentives to

[2] L. Rafael Reif, "The Innovation Deficit," *Boston Globe*, December 29, 2013, K8.

[3] "Companies Reshoring," *Reshoring Initiative*, accessed February 19, 2017, http://reshorenow.org/companies-reshoring/.

its suppliers to source products in the United States.[4] Whether this "buy America" preference will grow or taper off is hard to predict. What is clear, however, is that when firms decide to reshore, the jobs that return to or remain in the United States are not the same as the ones that were outsourced. They tend to be fewer in number and to be changed in some way that offsets the lower wage advantages of other countries. In some cases, employers that reshore jobs require U.S. workers to accept lower wages than were paid when the jobs were originally outsourced. In other cases, the jobs are redesigned to complement the use of advanced technologies. Although there are fewer of them, they tend to exhibit many elements of the high-road approach.

Local economic development agencies often compete for and subsidize these decisions about where to locate a company, and the incentives they offer can influence the choices employers make. Some offer tax breaks; some market their region as low wage and union-free; some focus on providing good technical school and other training programs. Most emphasize a mix of all these incentives. Thus, while reshoring in general is a positive and welcome development, it is essential that local governments choose incentives that will influence employers to create high-quality jobs. This requires local communities to develop what we described in Chapter 5 as an "educational ecosystem" of workforce training and development institutions that are capable of supplying workers with the state-of-the-art skills they need for modern manufacturing. Having such educational ecosystems in place would go a long way toward creating a strong incentive for firms to locate and pursue high-road strategies that translate into high quality jobs.

While there may be other ways to bring back jobs that have been lost such as changing tax rules on repatriated profits or strengthening the enforcement of anti-dumping regulations (laws that keep foreign firms from selling goods in the United States at below their actual cost), the long-term viability of U.S. manufacturing will likely depend on whether the United States can be a leader in developing next-generation manufacturing industries. The federal government has provided research and development funding to create advanced manufacturing institutes in about a dozen industries. Most of these involve five-year funding commitments. To date, these projects have focused mainly on research and on developing the technologies that will support these new industries. Each institute has a workforce development component that is charged with preparing the industry's next-generation work force.

[4] Cindi Marsiglio, "Walmart's Investment in U.S. Manufacturing, Explained," *Walmart Today*, June 16, 2016, accessed February 19, 2017, http://blog.walmart.com/us-manufacturing/20160616/walmarts-investment-in-us-manufacturing-explained.

A review of the status of these programs by the consulting firm DeLoitte raised concerns that insufficient attention is being given to integrating job and work design considerations directly into the technology choices and design processes that are under way. Recall Vivienne Ming's admonition about AI: "We need technologists to think about how their work will have reverberating effects on society. . . . These should be collective decisions we make as a society because they have broad societal implications." Dr. Ming's point needs to be built into the design processes and the decisions about where to locate manufacturing in these new industries. Given the fact that federal government funds are helping to jumpstart these industries, it is fair to hold firms that receive funding accountable for creating high-quality jobs and locating them in the United States. Our bottom-line recommendation is this: The federal government should continue to fund advanced manufacturing institutes but should increase the standards related to how to best design and create high-quality jobs and make those considerations an integral component of research and development. The government should also support efforts of industry and labor leaders to match training and education strategies to the job requirements of these advanced industries.

But we should also be realistic. New manufacturing industries will be highly automated. There is no good way to estimate the number of new jobs they will generate, but they will not come anywhere close to approximating the number of manufacturing jobs in the United States of the past. They will, however, have positive "multiplier" effects: For every new manufacturing job created, some multiple number of supplier and service jobs will also be supported. Investments in research and development are likely to have equivalent if not greater multiplier effects as they spawn further innovations and new applications for the technologies developed. This, in the end, is one of the key reasons why having a strong manufacturing base in America is so important, regardless of the number of direct manufacturing jobs that are created.

In summary, the first order of business for the federal government is to make the investments needed to support economic growth, basic science, and technological innovation that lay the foundation for a globally competitive, innovation-driven economy. But macroeconomic strategies alone are not enough. The legacy of work and employment policies and enforcement models that have carried over from the industrial economy of the twentieth century also need to be updated to better match today's economy, work force, and work arrangements.

Bringing Employment Policies into the Twenty-First Century

Recall that Frances Perkins told President Roosevelt that if he hired her as his secretary of labor, she would fight for the policies she wanted. When he gave her the green light to pursue that agenda, she turned to a cadre of experts who had direct experience in designing and working with state and local innovations that served as the model for the New Deal. We will need an equivalent infusion of expertise and experience to advise President Trump and push Congress to pass a set of new employment policies and enforcement strategies that are tailored to the needs of the next generation and the economy. The good news is that a large army of academic, industry, labor, and community professionals has built up this knowledge and experience. Now is the time to put it to work!

Educational Ecosystems for Middle-Skill Workers

A good place to start would be to build on the large body of research and experience that exists in the area of work force training and development. There is clear recognition that despite the effects of technology, America will continue to have a need for "middle skilled" workers to replace the retiring baby boom generation of technicians, mechanics, skilled trades workers, logistic specialists, and skilled service workers in the utilities, communications, and entertainment industries. A great deal has been learned over the years about the factors that are critical to the success of technical training programs, and the United States has a strong infrastructure of vocational schools and community colleges and apprenticeship training programs to build on in meeting these needs. What is required at this point, however, is strong federal support and coordination of these efforts and firm commitments from business, labor, and education leaders to work together to bring the programs to their full potential.

Support and Flexibility for Working Families

In Chapter 1 we noted that the United States has the distinction of being the last holdout against a national policy that supports working parents with some form of paid family leave. In the absence of a national policy, a number of states, such as New Jersey, California, Rhode Island, and New York, have taken the lead in enacting forms of paid family leave. California enacted its law in 2004. A 2010 follow-up study by Eileen Appelbaum and Ruth Milkman on the effects of the new benefit reported these findings:

> PFL [Paid Family Leave] use is associated with better economic, social, and health outcomes for workers and their families. . . . Workers in low-quality jobs who used PFL were more likely than those who did not use it to return to the same employer after a family leave, were more satisfied with the length of their leaves, were better able to care for newborns, and were better able to make childcare arrangements.

> The study found that despite strong opposition to passage of the PFL bill by employers, five years after implementation, the new PFL program was a "nonevent" for the vast majority of businesses. Our data reveal that employers themselves report that the PFL had no effect or a positive effect on the productivity, profitability or performance, turnover, and worker morale of their organizations. Moreover[,] many employers enjoyed cost savings as a result of the program, and abuse was rare.[5]

One limitation of the California program, however, is that many low-income workers were unaware of it or were afraid that using it would jeopardize their future employment. The implication is that even with legislation like this, community, immigrant, labor, and other groups need to educate workers about gaining access to benefits they are entitled to use and support them in their efforts to do so.

A growing number of states and localities (as of this writing, seven states, twenty-nine cities, two counties and the District of Columbia)[6] have taken a slightly different approach to supporting working families by enacting paid sick leave that can be used both for personal illnesses and for time needed to care for a family member. The growing popularity of these local initiatives again reflects the pressures parents experience in meeting their dual work and family responsibilities. Federal legislation would be the next logical step in this direction.

There is also growing attention to flexible work scheduling and related practices among private sector firms.[7] To date, most of the attention has been focused on flexibility for professional and white-collar employees. However, labor advocates have helped raise the profile of scheduling practices for service sector employees in

[5] Eileen Appelbaum and Ruth Milkman, *Unfinished Business: Paid Family Leave in California and the Future of U.S. Work-Family Policy* (Ithaca, NY: Cornell University/ILR Press, Press, 2010). For another study of the effects of this law that reported similar findings, see Charles J. Baum and Christopher J. Ruhn, "The Effects of Paid Family Leave on Labor Market Outcomes," National Bureau of Economic Research Working Paper 19741, December 2013, accessed February 19, 2017, http://www.nber.org/papers/w19741.pdf.

[6] "Paid Sick Time Legislative Successes," *A Better Balance*, updated January 4, 2017, accessed February 19, 2017, http://www.abetterbalance.org/web/images/stories/Documents/sickdays/factsheet/PSDchart.pdf.

[7] See Lotte Bailyn, *Breaking the Mold: Redesigning Work for Productive and Satisfying Lives*, 2nd ed. (Ithaca, NY: Cornell University/ILR Press, 2006); and Erin L. Kelly, Phyllis Moen, and Eric Tranby, "Changing Workplaces to Reduce Work-Family Conflict: Schedule Control in a White-Collar Organization," *American Sociological Review* 76, no. 2 (2011): 265–290.

firms such as Starbucks. This is likely to be an active arena for innovation in the years ahead.

Evaluations of the initiatives taken by state and local governments and by private-sector firms clearly have demonstrated the viability and value of flexible work and family programs and practices. Now is the time to move these efforts to the national level. Putting them at the top of the employment policy priority list would send a strong signal that the federal government is prepared to take action that is consistent with today's work force and today's economic realities instead of continuing the outdated and out-of-touch debates that repeat past battles between traditional labor and business groups.

Moving from Firm-Based to Portable Benefits

The old social contract was based on a long-term employment relationship with an employer that provided an extensive array of benefits that included health insurance and a pension plan. Today's employment relationships are more diverse and more uncertain in duration, and they increasingly either shift the costs of these benefits to employees or fail to provide them at all.

Firm-specific benefits grew out the work of the World War II–era National War Labor Board. The board encouraged firms to provide or expand health and retirement benefits in lieu of granting higher wage increases. That strategy worked brilliantly as long as workers stayed with a single employer for most of their careers. Clearly those days are over. But alternatives are not yet in place that are as effective in ensuring against the risk of individual or family illness or injury or that would help workers save for a secure retirement income.

The enactment of the Affordable Care Act in 2010 was the first major step toward weaning ourselves away from employers as the primary providers of health insurance. To date, the new policy has provided health insurance coverage to 20 million people who were previously uninsured. But it has generated considerable controversy among those who oppose a strong government role in health care insurance. It clearly is not the final solution to meeting the nation's objectives for an efficient, cost effective, high quality, and accessible health care system. Yet it is clear that returning to health care plans provided by individual employers is no solution either. This point was brought home at an annual meeting of the National Academy of Human Resources, which brings together the top human resources executives in the country. How Obamacare would affect their firms in the short run and over the course of the decade was a major topic at the meeting. Two points of consensus emerged from the discussion: (1) the companies meeting participants work for had no plans to improve their health care benefit offerings—all were looking for strategies to further cut costs or shift a higher percentage of the costs to current or retired employees; and

(2) these executives predicted that by the end of this decade their firms would be out of the health insurance business altogether. Unfortunately, none of the executives suggested a plan for how to turn their predictions into reality in that time frame.

The retirement issue is more dire. Social Security, the baseline, near-universal retirement system, will need some adjustments if it is to stay solvent for the next generation. This is not an insurmountable challenge—some combination of relaxing the income limits on social security contributions, increasing the eligibility age modestly to account for longer life expectancies, and/or modifying the cost of living or some other aspect of the benefit formula can ensure long-term solvency. This will take political courage, to be sure—after all, Social Security is not called the third rail of politics for nothing.

A bigger problem than making Social Security solvent is decline in employer-provided private pensions. The pension system that emerged during and after World War II and was a key part of the old social contract is dying a slow but steady death. At its highest point, in the 1970s, it covered just under half of the labor force. Now only about 20 percent of the work force has some form of defined benefit plan. Most of these are public-sector employees.[8] Many private employers have dumped pensions plans, and the 401ks and other forms of defined contribution plans that have replaced them are not filling the retirement security gap. As we saw during the Great Recession, these funds are subject to higher risk of losing value when economic growth slows or declines.

The effect of this is to reduce the incentive for individuals to retire just when we most want them to —when the economy is in trouble and young people are having the most difficulty finding a job. In short, 401ks are not a substitute for a real retirement saving plan. They are a useful supplement, but they have not and will not serve as an adequate retirement program for the next generation. Something else is needed.

One simple solution has been proposed by Teresa Ghilarducci, a faculty member at the New School for Social Research. She outlines it in a book with the catchy title of *When I'm Sixty-Four*.[9] Her plan calls for employees to contribute 5 percent of their income, part of which could be paid by their employer. The savings would be put into a government-administered fund and the government would guarantee a minimum annual rate of return of 3 percent on the invested funds that

[8] "Employee Benefits Survey," U.S. Department of Labor, March 2016, https://www.bls.gov/ncs/ebs/benefits/2016/benefits_retirement.htm.

[9] Teresa Ghilarducci, *When I'm Sixty-Four: The Plot Against Pensions and the Plan to Save Them* (Princeton, NJ: Princeton University Press, 2008).

would then turn into a lifetime annuity at retirement age. This proposal deals with many of the problems associated with the decline of traditional pension plans and the inadequate 401ks or similar plans. Professor Ghilarducci's plan does not depend on individual employers to provide a pension program, it does not shift all the risk of retirement savings to individuals, it does not create the perverse cyclical retirement disincentive, and, like Social Security, it is universal in coverage regardless of the type of employment relationship (including independent contractors) one has at various points in his or her career.

This is one well-thought-out option for meeting the retirement security needs of the next generation. Others might be invented. Or we could take an even easier route by simply expanding coverage and increasing the benefit levels of Social Security. The task today is to squarely face the need for a stable, universal, and adequate supplement to Social Security or an increase in Social Security retirement benefits. The baby-boom generation benefited from the decisive actions President Roosevelt and his advisors took in 1935. We need to take a similar action now or the baby boomers who benefited from Roosevelt's actions had better be prepared to subsidize their children's retirement or leave many of them to retire into poverty.

From Responsible Employers to the Fissured Workplace

In the heyday of the old social contract, most people could tell you who they worked for and have no doubts about it. They worked for GM or IBM or the local department store, gas station, insurance company, hotel, utility company, and so on. Today ask someone who helps build cell towers or works in a big brand-name hotel or services your home Internet connection who he or she works for and you are likely to get a wide variety of answers. And if you ask their employer who is responsible for enforcing core labor standards related to workplace safety and health, the payment of wages and overtime rules, and equal employment opportunity, you are likely to get even more ambiguous answers. This diffusion of responsibility has created a global economy where bad things can happen because no single regulatory agency oversees the business world. We saw this with the Gulf Coast oil spill in 2010. British Petroleum (the owners of the well with rights to the oil that spilled), Halliburton (the contractor that provided testing services related to the stability of the drilling process), and Transocean (the company that provided the drilling platform) each pointed the finger at the others in an effort to shirk responsibility for the accident and the financial liabilities it generated.

David Weil, the former administrator of the Wage and Hour Division of the U.S. Department of Labor, is the nation's leading academic expert on what he calls the "fissured workplace" and how to update enforcement of basic employment standards to adapt to these new business realities.[10] He notes that for most hotels, the name on the marquee may not be the company that owns the property, a different company may manage the specific facility, and many of the tasks related to running the hotel such as housecleaning may be contracted out to yet another company. Moreover, these properties and management contracts change periodically so that the mix of employers is in constant flux.

Weil and others who have studied this issue argue that changes in both legislation and enforcement strategies are needed to adapt to these fissured employment relationships. It is clear how the laws need to change. New laws are already being tested in a few states such as California, New York, and Illinois. The idea is simple and direct: Hold the firm that purchases the ultimate goods or services or sets the standards for the work responsible for assuring that basic employment laws and standards are met by all firms and contractors that contribute to the product or service. In the case of the Gulf Oil disaster, British Petroleum would have been the responsible employer. If a policy of holding the ultimate purchaser of goods and services responsible had been in place, British Petroleum likely would have taken more care to ensure that its managers controlled key decisions. Because such a policy was not in place, some of these decisions fell through the cracks of shared or ambiguous authority and reporting relationships. Investigators later found that the absence of clear channels of responsibility was the root cause of the accident. The same assignment of responsibility would be true of the giant oil and chemical companies that subcontract some of the most dangerous maintenance work when they retool parts of their refineries and processing plants. The same assignment of responsibility would be true of AT&T and other providers of cell phone service when they issue contracts and subcontracts to specialized firms that build their cell towers. And the same would be true when Hilton or Marriott or Starwood contracts with others to manage a hotel property bearing their company's name.

The enforcement strategy Weil endorses relies on a mix of carrots and sticks. It would reward firms that have high standards and workplace practices that achieve high compliance (that is, low violation rates) and would target enforcement resources on the most egregious violators of employment standards. Making this strategy work will require better and more coordinated information gathering and sharing of knowledge of violation and compliance rates across different government agencies such as the Occupational Safety and Health Administration (OSHA), the Equal Employment Opportunity Commission (EEOC), the National Labor Relations Board (NLRB), and the Wage and Hour Division of the Labor Department—the alphabet

[10] David Weil, *The Fissured Workplace* (Cambridge, MA: Harvard University Press, 2014).

soup of quasi-independent government agencies that enforce different labor standards. It would also require these agencies to work with and empower community groups, unions and professional associations, immigrant organizations, and individual workers to report violations, complement the work of enforcement agencies by monitoring workplace practices, and work cooperatively with employers to ensure compliance and gradually upgrade standards so their workplaces can move from the targets of regulatory agencies to the rewarded side of the enforcement ledger. President Obama signed the Safe and Fair Workplace Order, an executive order that would require private sector firms to report their employment law compliance/violation records However, this order has not yet been implemented. Doing so would be a step in the right direction.

The government also needs to model high-road business practices. In December 2013, a *New York Times* front-page headline announced that the agencies of the federal government are among the biggest purchasers of goods made in sweatshops around the world.[11] The president needs to assign responsibility to a single high-level official for creating, monitoring, and enforcing the government's own code of conduct throughout its supply chain. It is time for the government to learn from the work that has already been done by private firms and the multi-stakeholder groups that are actively addressing this issue in global supply chains.

In 1965, President Lyndon Johnson signed an executive order that required all federal contractors to demonstrate that they were taking affirmative action to end discrimination in their employment practices. An equivalent executive order that would require all federal contractors to demonstrate that they are taking affirmative action to enforce the full array of U.S. labor standards and pay fair wages would go a long way toward supporting and broadly diffusing the high-road strategies needed to support the next-generation social contract. Going a step further, incentives could be included in government contracts that reward firms that have high-road strategies and employment practices in place. This would be the carrot side of a carrot-and-stick strategy for leveling the playing field between high- and low-road firms.

Independent Contractors and the "Gig" or "Platform" Economy

A lively debate is under way among researchers and statisticians over the current and future number of workers in the so-called "gig" or platform economy (think Uber

[11] Ian Urbina, "The U.S. Flouts Its Own Advice in Procuring Overseas Clothing," *New York Times*, December 22, 2013, accessed February 19, 2017, http://www.nytimes.com/2013/12/23/world/americas/buying-overseas-clothing-us-flouts-its-own-advice.html.

drivers, workers who obtain jobs through Shiftgig, service providers hired through TaskRabbit, etc.) The most reliable survey completed to date put the number at less than 1 percent of the overall work force[12] but the number is growing rapidly and some even expect these workers to constitute a substantial portion of the work force at some point. The prospect that this category will continue to grow has generated lively debate about whether and, more realistically, how independent contractors should be covered by the basic employment standards that protect regular employees. (Recall that independent contractors are not protected by federal wage and hour laws, equal employment opportunity laws, or labor law.)

A number of options have been proposed for dealing with this issue. The U.S. Wage and Hour Division has issued guidelines to reclassify employees who have been incorrectly misclassified as independent contractors.[13] Workers in California and elsewhere have filed lawsuits, and state regulatory agencies are reviewing this issue for Uber drivers and other groups. It is important to classify such workers as employees because that status would entitle them to the protection of federal and state labor laws and possibly to benefits from their employers.

But clarifying the status of these groups of workers as employees would not solve the issues those who are legitimately classified as independent contractors face. As one way of dealing with those issues, some have proposed making benefits such as unemployment insurance, retirement benefits, and other protections portable for independent contractors. Each platform provider would be responsible for funding these benefits in proportion to the hours they "employ" the independent contractor.[14] Another suggestion is to create an intermediate classification of "dependent contractor" and to extend coverage of labor laws to contractors who work with platform companies. In this strategy, "dependent contractors" would have the right to organize and negotiate with platform providers in ways that fit the circumstances of this particular group.[15] Still others propose that independent contractors be allowed

[12] Lawrence F. Katz and Alan B. Krueger, "The Rise and Nature of Alternative Work Arrangements in the United States, 1995-2015," *National Bureau of Economic Research*, March 29, 2015, accessed February 19, 2017, https://krueger.princeton.edu/sites/default/files/akrueger/files/katz_krueger_cws_-_march_29_20165.pdf.

[13] "Misclassification of Employees as Independent Contractors," *U.S. Department of Labor*, accessed February 19, 2017, https://www.dol.gov/whd/workers/misclassification/.

[14] David Rolf, Shelby Clark, and Corrie Watterson Bryant, "Portable Benefits in the 21st Century," *The Aspen Institute*, 2016, accessed February 19, 2017, https://www.aspeninstitute.org/publications/portable-benefits-21st-century/.

[15] Seth D. Harris and Alan B. Krueger, "A Proposal for Modernizing Employment Law for the Twenty-First-Century Work: The 'Independent Worker,'" The Hamilton Project Discussion Paper 2015-10,

to organize and be protected from discrimination. A European group of trade unions has developed a proposal for this that also favors developing more cooperative forms of consultation and information sharing similar to those now found in union-management partnerships.[16] Perhaps the most comprehensive approach has been suggested by Senator Elizabeth Warren. She proposes extending basic safety net and labor standards coverage and the right to organize to all who work while simultaneously encouraging worker organizations or other groups to take responsibility for administering benefits for workers as they move across different jobs.[17]

Thus, while a growing number of people are studying and debating the merits of these different options, no consensus has yet emerged. One difficulty in formulating workable policies is that the various employment and labor laws use different criteria to define the boundary between employees and independent contractors. Standardizing one common definition would be a good first step toward clarity for employers, government regulators, and the work force. For now, perhaps it is best to agree with the broad principle Senator Warren has espoused and encourage experimentation and evaluation of different ways to govern work in the contractor and gig sectors of the economy.

What about Trade, Globalization, and Jobs?

Perhaps no issue has been subjected to more controversy at this juncture than globalization and trade. The good news is many now recognize that globalization indeed has produced losers as well as winners and that this pattern is not sustainable without risking more backlash against it. The policies recommended above would go a long way toward providing job growth and opportunities for a broader cross-section of the work force to gain access to good-quality jobs. But the political backlash in the United States against the globalization of business will not fade away without further

December 2015, accessed February 19, 2017, http://www.hamiltonproject.org/assets/files/modernizing_labor_laws_for_twenty_first_century_work_krueger_harris.pdf.

[16] N.a., "Frankfurt Paper on Platform-Based Work: Proposals for Platform Operators, Clients, Policy Makers, Workers, and Worker Organizations," December 6, 2016, accessed February 19, 2017, https://www.igmetall.de/docs_20161214_Frankfurt_Paper_on_Platform_Based_Work_EN_b939ef89f7e5f3a639cd6a1a930feffd8f55cecb.pdf.

[17] Remarks by Senator Elizabeth Warren, "Strengthening the Basic Bargain for Workers in the Modern Economy," New America Annual Conference, May 19, 2016, accessed February 19, 2017, https://www.warren.senate.gov/files/documents/2016-5-19_Warren_New_America_Remarks.pdf.

direct actions to help the individuals, families, and communities hit hardest by the loss of good jobs through international trade, automation, or other developments. A direct policy of community redevelopment and income support may need to be part of the price of rebuilding support for sensible trade policies in an age of globalization.

What would such a policy look like? This is largely uncharted territory, so it might be best to experiment with multiple approaches before choosing one or more as the direction for national policy. One simple option that has gained growing support in recent years is wage insurance. That is, instead of assuming, as our original unemployment insurance model did, that laid-off employees would eventually return to work with their prior firms or find suitable substitutes in their local labor markets, perhaps we need to face the fact that some of those who have lost good jobs will never find equivalent employment. Instead, as one study showed, those who were laid off from manufacturing jobs and found replacement jobs tended to lose 15 to 20 percent in wages (and perhaps more in benefits). Many (but an as-of-yet unknown number) older workers never found jobs that came close to replacing the income from their previous jobs. A wage insurance approach sees this as a cost of globalization that should be absorbed and funded through tax policies aimed at higher-income individuals and businesses that benefit from the flexibility to lay off "redundant" workers. Public funds would be used to make up part of the difference between a laid-off worker's previous job and the wages paid in the replacement job he or she accepts. Numerous economists and policy analysts have proposed experiments to test the effects of wage insurance.[18] Careful review of this option is clearly warranted. If it is found to be successful, it could be part of a comprehensive strategy for compensating those who bear the most costs of adjustment to globalization and technological change.

A more extreme form of wage insurance is being promoted under the label of a guaranteed annual income. The idea is to give everyone an amount that would be equivalent to or approach the level of a basic living wage, perhaps in lieu of other welfare and transfer payments for specific needs (such as disability payments or food stamps). This idea has been floated by Silicon Valley executives and, among others, the highly regarded former labor leader Andrew Stern.[19] Those of us who worry about the multiple roles work plays in our lives and in society—its moral, psychological, social, and motivational benefits, not to mention economic features—are less than

[18] Stephen Wandner, "Wage Insurance as a Policy Option in the United States," unpublished paper, January 16, 2016, accessed February 19, 2017, http://www.upjohn.org/sites/default/files/pdf/wage-insurance.pdf.

[19] Andy Stern, *Raising the Floor: How a Universal Basic Income Can Renew our Economy and Rebuild the American Dream* (New York: Public Affairs, 2016).

enthralled with this idea. Some small-scale experiments with this approach are under way in Finland and Canada. Perhaps we can learn from these experiments and then let the debate over this approach proceed.

At the nadir of the Great Depression, the federal government invested in a variety of direct public service programs to create jobs and develop communities. Will we need to consider modern-day equivalents? Ideas such as one- or two-year mandatory or voluntary public service programs for young people fall into this category. We already have a number of such programs. Obviously, the largest is voluntary enlistment in the armed forces. Smaller programs such as the Peace Corps, AmeriCorps, and Teach For America all aim to provide options that enable young people to serve important community needs while gaining experience at jobs for which they receive very modest levels of compensation. Encouraging young people to join programs such as these as a bridge from school to their first career job would both serve a variety of important social needs and enhance and deepen the human capital of young people as they enter the labor force.

And Yes, We Do Need a New Labor Policy

Just as the National Labor Relations Act was the most contentious part of the New Deal legislation in the 1930s, so too are debates today over how to replace the nation's defunct labor law. Let there be no doubt: the old National Labor Relations Act is "ossified," as labor law expert Cynthia Estlund has so accurately described it.[20] For years, the law has failed to deliver on its promise of providing workers access to collective bargaining, as a national commission noted twenty years ago. The clearest evidence that the law is dysfunctional was summarized in Chapter 6 but is worth repeating here: Less than one in ten organizing efforts succeeds in getting a labor contract if management opposes the organizing effort. This is the case even when a majority of workers indicate that they want to be represented by a union. The bottom line is that today any employer who wants to oppose efforts to organize a union through legal or illegal means has about a 90 percent chance of succeeding. That means that it is no longer workers who decide whether they will have an independent voice at work. Lawyers, consultants, and their employers control this decision. That was not the intent of the National Labor Relations Act when it was enacted!

The problem is even deeper. The many adversarial doctrines, traditions, and norms that built up in traditional collective bargaining over the last eighty years no longer produce a competitive or high-quality workplace. The high-road workplaces discussed throughout this book are either nonunion or are unionized settings in

[20] Cynthia Estlund, *Regoverning the Workplace* (New Haven, CT: Yale University Press, 2010).

which management and labor leaders work together in partnerships that ignore or put aside many (if not all) of the old doctrines and traditions. Adversarial labor-management relationships can no longer compete with these alternatives.

A modern labor policy therefore must have two equally important objectives: 1) to restore the ability of workers to gain an independent and effective voice at work; and 2) to promote collaborative forms of worker voice and engagement that serve the mutual interests of the work force, employers, and the clients, patients, students, or other ultimate customers they serve. And given the emerging innovations with different forms of work, labor policy needs to open up to an era of experimentation with, evaluation of, and ultimately endorsement of those new forms of worker voice that demonstrate their value to different groups of workers, employees and independent contractors alike.

Here again the seeds of an innovative labor policy have been sown by the companies and unions that have implemented different types of partnerships and by worker entrepreneurs inside and outside the labor movement. Saturn and Kaiser Permanente are the most comprehensive and perhaps most complex labor-management partnership models. Ford and the UAW have fashioned an approach that works for them, as have Southwest Airlines and its unions. A number of hotels and the workers' organization UNITE HERE have reached agreements on the partnership approach in their industry. The key, however, lies in government endorsement of these new approaches. It needs to open labor law and labor policy to more of this type of experimentation and to further support efforts to go outside the parameters of the National Labor Relations Act. This will require leadership and outreach by the government's chief labor officer—the secretary of labor.

But there are those would argue that labor unions and other forms of collective worker voice are dinosaurs and/or are no longer needed. They claim that competitive labor markets and benevolent employers will serve as "labor's voice." While this debate will likely be front and center in the years ahead, we hope it will be joined by careful analysis of both the history of labor's decline and more aggressive efforts to suppress collective forms of voice and by evaluation of the forms of voice now emerging from the work force. Ultimately, workers' preferences should rule in a democratic society.

Making It Happen: An Action Plan for Government Leaders

This is arguably a very ambitious and far-reaching agenda for transforming labor and employment policy. But anything less will likely fail to provide the equivalent of the New Deal platform on which the old social contract was built gradually over subsequent decades. While this agenda is ambitious, it is also entirely doable. The ideas and actions outlined here are all derived from experimentation that was done in the private sector or in state or local governments. Some come from the research and ideas of academics, but none come from ivory-towered theorists who are far removed from the practical world of work. The ideas come from evidence generated from field studies of what is working on a small scale.

Now is the time to test these on a scale large enough to meet the needs of the next generation. So here is an action plan federal policy makers can implement to do their part to support development of a new social contract.

- Carefully review and assess innovations in state and local governments on issues such as minimum wages and paid family and sick leave and determine how to adapt the most effective measures into a national policy. Learning from local innovations should be an ongoing process.

- Invest in the future: Take up the long overdue task of rebuilding America's infrastructure. Restore federal funding for science and engineering to its earlier levels. Continue to fund projects that are under way that are working to developing the next generation of manufacturing industries and strengthen the role of work design and workforce development within those initiatives.

- Embrace the realities of the global economy but take steps to close the gaps between those who benefit and those who are absorbing the costs of globalization with active labor market and community redevelopment policies and, if necessary, income subsidies.

- Support high-road business strategies and workplace practices. Raise the floor and enforce basic labor standards to nudge low-road firms to move upward and reward the presence of high-productivity, high-wage practices in firms bidding for government contracts.

- Stay the course in weaning basic benefits such as health insurance and pensions from individual firms. Transform benefits to be portable so they move with workers throughout their career, regardless of who they work for or whether they are classified as employees or contractors.

- Apply the lessons learned about effective ecosystems for workforce training and development. Hold vocational schools, community colleges, apprenticeship programs, and other training institutions accountable for meeting the standard of engaging with multiple employers, mixing on-the-job and classroom training, and placing graduates in jobs with career ladders that support lifelong learning.

- Break the long-standing impasse over labor policy by redefining the goals of labor policy to protect workers' rights to organize, promote effective models of labor-management partnerships, and encourage and support experimentation with new forms of voice suited to a more mobile and diverse work force.
- Finally, and most important, lead the effort to bring the stakeholders together to restore respect for and dialogue with each other and work to achieve a new social contract suited to today's economy and work force.

Video Resources

The following video covering topics in this chapter can be found at http://iwer.mit.edu/speak-up-for-work/:

Will Government Pave or Block the Path to a New Social Contract?

8

A Call to Action

Now we come to the hard part. We have outlined what research, experience, and emerging innovations tell us that business, labor, education, and government leaders and the work force itself can do to build a new social contract. But we have also said that none of these groups can make this happen on their own. The question that remains is how might they work together to make it happen. We take up this question here.

First, let's put all of this in context. We write at a time of deep divisions in society and considerable anxiety and uncertainty about the future of the economy and democracy in the United States and around the world. Populist political candidates are gaining strength across Europe. The British vote to leave the European Union pitted young against old and thriving Londoners against those living in struggling regions and small towns. A similar voting pattern put Donald Trump in the White House. He lost in thriving states and cities and won in states and regions hardest hit by job losses. Black and Hispanics and whites with college degrees voted one way, white blue-collar workers voted another, and a relatively large number of people chose not to vote at all.

On the jobs front, those with the skills to match industry's thirst for "big data" or filling advanced manufacturing and technical jobs are thriving; those laid off from manufacturing jobs that have been automated or shipped overseas have likely taken permanent cuts in pay and benefit or have moved to the margins of the labor force or out of it entirely. Many in low-wage service sector jobs do not make a living wage. And many in the emerging platform or on-demand economy work without the protections of basic labor standards or benefits.

This is either worst or the best of times to call for efforts to come together to begin working on building a new social contract. The need is clear and urgent. The divides that need to be closed are as great as any we have seen since we entered this field of work nearly half a century ago. Moreover, the divides in the labor market long preceded both the current political developments and the Great Recession. Perhaps it had to get this bad before it could get better. One thing is clear—these divides cannot continue. We may just be seeing the first wave of anger and political action they have produced. Now it is time to channel that anger into positive actions to deal with their root causes. Our purpose in this chapter is to lay out a path forward that we believe will help heal the economic and political divides. It calls for all hands on deck.

Students Show Us the Way

Let's begin with an illustration of what we have in mind. As part of our online and on-campus classes we have been conducting a social contract exercise to simulate the multiparty dialogue and coordinated effort that will be needed to put the economy back on a more inclusive track. We present the most recent results from the exercise not because the students provide all the answers to what needs to be done but to illustrate the potential for such an exercise among the real players who can make a difference.

The exercise has two phases. In phase 1, participants are given a list of potential issues that each stakeholder (business, labor, education, and government) could address to help build a new social contract. The issues are categorized into five broad topics—Workforce Capabilities, Fair Treatment/Worker Representation, Organizational/Enterprise Performance, Rewards/Benefits for Workers, and Family/Community/Society. The students choose the top two issues they want each stakeholder to address and then write out specific proposals for what they want that stakeholder to do about those issues. Our teaching staff (with the help of MBA students) then reviews the issues that get the most proposals within the different topics for each stakeholder and summarizes them as consolidated proposals for phase 2 of the exercise. In phase 2, class members are asked to rank the priorities they would attach to these consolidated proposals and to rate their overall satisfaction with the new social contract that would be produced if each of the proposals were to be implemented.

Figure 8.1 Stakeholder Priorities for Different Topics

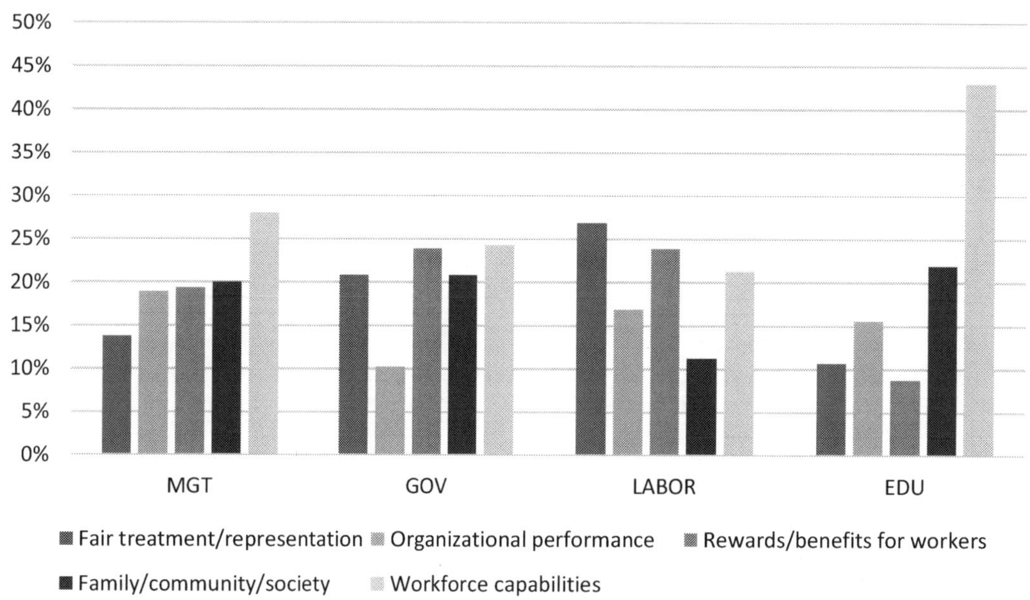

Figure 8.1 summarizes the results from the phase 1 survey showing the priorities the students assigned to each of these topics for the different stakeholder groups. The data in the figure reveals some common patterns across the stakeholder groups and the specific proposals students offered to elaborate on what they expect the different stakeholders to do about these topics. All of the stakeholders gave high ratings to Building workforce capabilities through entry-level and ongoing training and development. There are also some obvious and predictable differences across groups. Issues such as health care benefits, parental leave, and equitable distribution of wages between workers and managers/executives contribute to the ratings on fair treatment and rewards/benefits for the business, labor, and government stakeholder groups. Students also expect government to address fairness by increasing minimum wages, updating labor and employment policies, and ensuring equitable access to education and training opportunities. They expect labor (both current unions and newer forms of worker advocates) to prioritize fair treatment through negotiating for wages, profit sharing or other forms of sharing gains from firm-wide or economy-wide performance, and assuring universal and portable health care.

Figure 8.2. Top Priority Social Contract Proposals for Each Stakeholder

Stakeholder	Highest-Priority Issue	Proposal to Address Issue
Management	Rewards & benefits for workers	All employees should expect to be paid a living wage and have access to affordable and portable healthcare options. Compensation systems should provide some means (profit-sharing, stock ownership, etc.) for all employees to share in the performance of the business. A dialogue in each company should be initiated over fair ratios of executive compensation to compensation for managers, supervisors, and lower-level workers.
Labor	Rewards & benefits for workers	Labor leaders need to continue to be strong advocates for living wages and access to healthcare for all workers. They must also encourage pay practices that link pay growth to the performance

		of the organization and the overall economy and negotiate for more equitable distribution of company profits across management and employees.
Education	Workforce capabilities	Educational institutions should teach skills that will prepare students to be successful in twenty-first-century labor markets. They should do this by partnering with businesses to ensure a practical and up-to-date learning experience. In addition, educational institutions should expand accessibility and options for lifelong learning (e.g., courses designed for mature workers).
Government	Rewards & benefits for workers	Governments should ensure that all workers (e.g., employees and contractors) receive portable and affordable healthcare benefits and minimum wages that are linked to increases in cost of living. Governments should also play an active role in the more equitable distribution of company profits between management and workers.

Figure 8.2 provides examples of the top priority proposals consolidated from phase 1 that illustrate the actions the class expects these stakeholders to take to contribute to a new social contract. A full list of these proposals can be found on our website (http://iwer.mit.edu). Here we report only proposals that received the top rankings from the class in phase 2 of the exercise. Note that in this second round, rewards and benefits get the top ranking for business, labor, and government while workforce capabilities remain the top priority for education (other stakeholders ranked workforce capabilities second or third).

These ratings and rankings provide only a snapshot of the expectations our class has for each of these stakeholders. But the messages embedded in some of the proposals offered by individual students further illustrate how class members want these parties to be more proactive. Here's a brief sampling:

> Government leaders should more strongly enforce basic labor standards to push low-road companies in the right direction. By doing so, workers will have better working conditions, which will therefore increase satisfaction, and boost productivity and the quality of the job.
>
> Labor leaders should expand the scope of labor unions to . . . not only help in their image building among the public but also become a source of rendering new skills to their members.
>
> Employers should reduce the portion of executive stock options in their executive compensation packages. Reducing Wall Street's pressure to maximize short-term profit would focus employers on investing in human and nonhuman capital instead of stock buybacks.
>
> Education leaders need to develop programs that are committed to lifelong learning. Public universities should work with major businesses in their communities to offer joint-training programs targeted at adults who need to develop their core skills. . . . Courses and learning must be contextualized wherever possible and have real-world applications.

Overall, the class gave a satisfaction rating of 8.0 on a 10-point scale for the complete social contract they generated. This suggests that it is possible, at least with this group of online students, to generate a new social contract that gains broad support. However, the less-than-perfect rating also suggests that there is still room for improvement.

The results of this simulation illustrate the possibility that leaders of these diverse stakeholder interests could forge constructive agreements through engagement with other stakeholders, informed by knowledge of today's realities and information about priorities and viable options. This is just a classroom exercise. But perhaps it could serve as a learning exercise for actual representatives of these key groups. If you want to give it a try, we have put two versions of the exercise on our website (http://iwer.mit.edu) that supports a negotiation among all the stakeholders. You can complete this exercise on your own to see how your views on what each stakeholder should do compares to the views of others who have participated in the exercise.

Another View: JUST Capital's National Work Force Surveys

Another body of data helps inform us about what the American work force expects of one key stakeholder group: American business leaders. One of the most comprehensive and data-driven efforts to assess whether firms are responsive to what the public expects of a good company is JUST Capital. Started by Paul Tudor, a highly successful former hedge fund manager, JUST Capital has collected a huge amount of data on the top 1,000 publicly traded companies and ranks them on criteria that *the public* says influences its judgment of a "just" company. The criteria, generated through a series of large-scale nationally representative surveys of American households, are shown in Figure 8.3. According to JUST's surveys, the top three criteria the American public uses to judge companies are whether they pay fair wages and benefits, whether they treat employees well, and the quality of the company's leadership and ethics. This is a clear message to business leaders. Perhaps the next step for JUST Capital is to collect the same data on what the public expects of the other key stakeholders!

Figure 8.3. Rankings of How the American Public Judges a "Just" Company

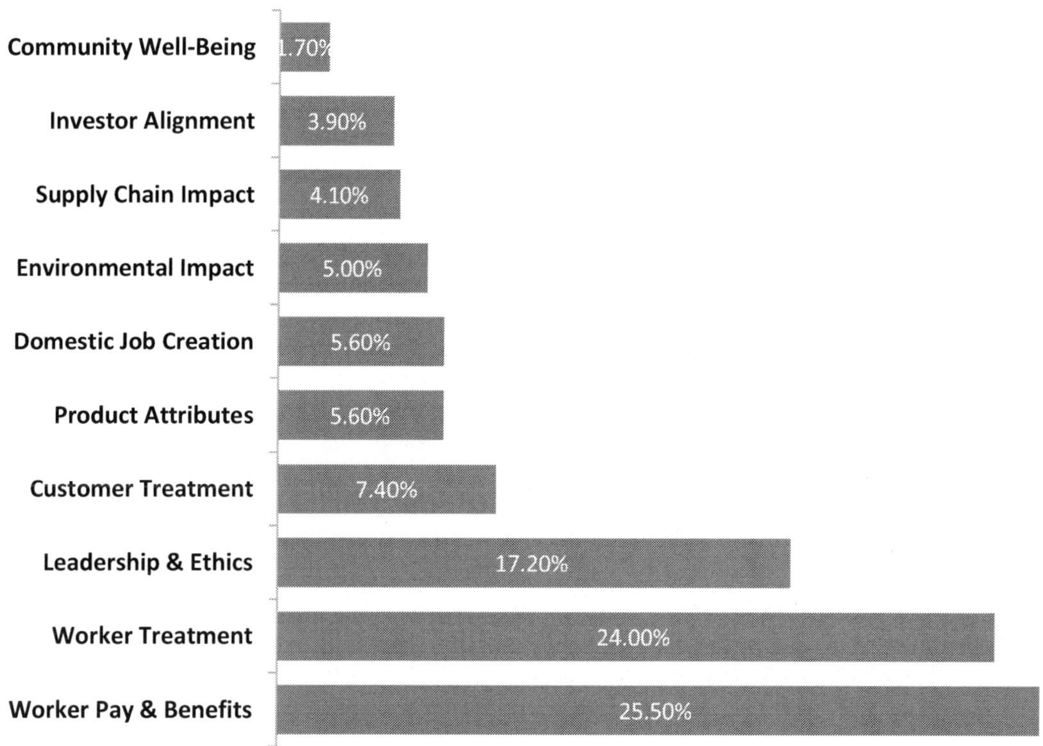

Source: JUST Capital 2015 Survey, http://justcapital.com/methodology/.

We now turn from the world of simulation and surveys to the real world as we suggest what actual leaders of these groups might do with others to forge a new social contract that can support good jobs and careers and put the economy on a path of sustained, shared prosperity.

Business Leaders

One of us participated in a recent gathering of CEOs in Europe to discuss the future of work. The meeting organizers chose to begin the meeting with a slide depicting business leaders and their societies sitting on a "tinder box" fueled by the pressures of movements toward political extremes, the pent-up anger of citizens over income inequality, the loss of jobs due to global trade, the need to absorb immigrants from war-torn countries, and fears that digital technologies on the horizon would eliminate more jobs than they will create. The "tinder box" metaphor resonated so well with these CEOs' fears for the future of business in their respective countries that they spent most of the meeting discussing what to do about it. The meeting ended with a collective commitment to work together and with other key stakeholders to address these pressures before they ignite and explode.

Why do business leaders need other stakeholders? Why can't we simply leave it to their individual self-interest to pursue the high-road strategies outlined in Chapter 4? Answering this question requires a discussion of the difference between the interests and responsibilities of individual firms and the collective interests and responsibilities of business in an advanced economy that is part of a democratic society.

First and foremost, individual business leaders are responsible for managing their organizations in ways that will achieve good profits. But as business leaders, citizens, and parents, they also want to provide good jobs and careers for future generations. They also have a collective economic interest in ensuring that American consumers have adequate purchasing power; even U.S. multinational firms still earn more than half of their revenue from domestic customers.

Throughout this book we have stressed that it takes a mix of well-crafted high-road strategies and work practices to achieve both sets of objectives. We also have emphasized that individual business leaders face a choice about whether to adopt a high-road strategy. While there is ample evidence that this approach can and does generate good financial returns for many individual firms, there is no guarantee that this approach is inevitably a better or superior way to maximize shareholder value than the alternative strategy of minimizing wages, tightly controlling the work force, replacing workers wherever possible either with new technology or by outsourcing noncore work to lower-cost domestic or global suppliers. So individual business leaders may not view it to be in the interest of their firm or in their own self-interest to invest in high-road strategies and then stay the course. Indeed, we highlighted a number of countervailing pressures that keep many business executives focused on low-road strategies such as pressures from Wall Street for short-term performance, the weakness of workers' voice and bargaining power as a counter to Wall Street pressures, outdated labor and employment policies, and financial incentives in compensation packages that keep executives focused on short-term results.

But what is in the interest of the individual firm may not be in the collective interests of business or society. Consider, for example, the comment in Box 8.1 from Ralph Gomory, the former chief scientist for IBM and the retired president of the Alfred P. Sloan Foundation. Gomory is one of the most thoughtful and credible critics of the narrowing of corporate goals that has taken place over the past several decades. He speaks from experience. He helped build a long-term research and development program at IBM and then created and oversaw an industry competitiveness program at the Sloan Foundation. His experience and wisdom as a thoughtful business leader give his words great credibility.

Box 8.1. Divergence of Individual Business and National Interests

[Government and business leaders] do not realize that the corporate goal of profit maximization at all costs does not serve the interests of the nation. They do not realize that the fundamental goals of the country and of our companies have diverged. The sole focus on profit maximization, which leads to offshoring and holds down wages, does not serve the nation. . . . We must act to realign the goals of company and country.

Source: Ralph Gomory, "A Time for Action: Jobs, Prosperity and National Goals," *Huffington Post,* January 25, 2010.

Gomory joins the growing chorus of voices suggesting that the financialization model of the American corporation has run its course and needs to be replaced with a broader view and that the new model must be complemented by strong and effective institutions in both government and civil society. But no one should have any illusions about how difficult this will be. One of the most powerful CEOs in the financial services industry made the obvious counterpoint at a Competitiveness Summit held at the Harvard Business School in 2011. After listening to Harvard professor Michael Porter speak of the need for firms to support efforts to improve the competitiveness of U.S. businesses and the living standards of U.S. workers, this executive commented: "I hear what you are saying, but every time I stand up and say we need to take actions to strengthen the U.S. economy, my shareholders cut my neck off, reminding me we are a global company."

His point was not that he was personally opposed to taking a longer-run approach or to reinvesting in order to grow jobs and strengthen America's competitiveness. Instead, he spoke the truth about the countervailing pressures he is under from his shareholders and from the financial agents and institutions that speak for them. So the question is how can this transformation in corporate objectives and behavior be achieved?

The place to start is to realize this is a classic market failure problem. Focusing solely on maximizing short-term profits may work for individual firms, but it creates costs that society must absorb. In contrast, a process in which business owners worked together with other stakeholders to share these costs would generate mutual benefits. So the key lies in mobilizing business leaders to work with each other and with government officials and civil society leaders who share an interest in and responsibility for the incomes, purchasing power, and standard of living of the country's next generation.

Where to start? Perhaps the easiest place, and the one with the highest payoff, would be the types of regional business and education consortia that have demonstrated their ability to overcome the market failures that hold back individual firms from subsidizing their employees' education at community colleges or universities. As noted in Chapter 5, model programs that mobilize groups of employers, such as those in California, Wisconsin, and other states, have demonstrated that it is possible to meet the needs of local employers and at the same time provide career opportunities with good rates of return for both employers and workers.

Apprenticeships and other joint union-management training programs and infrastructure projects also provide good opportunities for employers and unions to work together. In Chapter 5, we also noted how employers and unions in the utility industry looked ahead and saw the demographic cliff coming a decade ago as its skilled labor force approached retirement age. The industry's joint training and apprenticeship program is now producing the next generation of technicians and repair workers. These workers can look forward to good-paying jobs and careers. When individual firms cooperate with each other to create such programs, everyone benefits. All of the cooperating companies have access to the skilled and highly trained employees they need to remain competitive in their industries. In addition, the return on their investment in recruiting and training is much greater than it would be if each firm tried to create such a program on its own.

The same potential exists for businesses that need employees with the higher-level technical knowledge and skills that colleges and universities can teach via cooperative arrangements, online courses, and sponsored degree programs. Numerous companies have worked with programs such as Northeastern University's Cooperative

Education and Career Development program[1] to generate a steady stream of college graduates and postgraduates with state-of-the-art technical and managerial education and direct experience with applying this knowledge to real problems in industry. Online courses make it possible to expand such programs on a national scale. Doing so will go a long way toward reducing the costs of education and supporting the lifelong learning needs of the next-generation work force and providing employers with people who have the skills they are looking for.

Clearly there is ample potential for business leaders to work with each other and with education and labor organizations to ensure that the next-generation work force will meet their needs and be prepared for the jobs of the future.

American business leaders seldom speak with a common voice in public policy affairs, except perhaps when it comes to opposing nearly all proposals for raising labor standards and strengthening labor laws. We are both members of the Society for Human Resource Management, the largest association of professionals in our field. All too often we are embarrassed by the kneejerk opposition this organization espouses when, as been the recent case, overtime rules are revised, proposals to increase minimum wages are advanced, or even modest rule updates such as requiring firms to post notices that workers have the right to form a union are being considered. It is time for HR professionals to return to their roots, when they took pride in asserting their professional judgment about how to best balance or advance the shared interests of employers and the work force. Instead, too many HR professionals have become what Frank Doyle, the former vice-president of HR at General Electric, described as "perfect agents" of the CEO.[2] Indeed, we expect that HR and other business leaders will be tested to see what approach they will take if government officials try to roll back these and other changes introduced in recent years. We can only hope that the business community will stand up against such efforts and focus on sensible strategies for leveling the playing field between low- and high-road firms while respecting the need for all workers to have access to a voice on their jobs, a living wage, and fair employment conditions. It is time for business leaders to tell their lobbyists and trade or professional associations stop letting the bottom feeders dominate their political strategies.

And then there is dealing with Wall Street. In Chapter 4, we quoted the advice of Scott Stern that entrepreneurs need to "pick their investors carefully" so that they will not be pressured by venture capitalists or others to focus solely on the short term. Perhaps the time has come for executives in established firms to do the same and start

[1] http://www.northeastern.edu/coop/.

[2] Frank Doyle, "GE's Doyle Urges HR to Embrace a World of Change," *Work in America Report* 18 (March 1993): 3.

pushing back on Wall Street investment analysts who either don't understand or care about high-road strategies. It is encouraging to see that funds that invest in firms with strong reputations on environmental and social justice issues and that also perform strongly are growing rapidly. It is also encouraging to see organizations such as JUST Capital providing public rankings of firms that, according to their metrics, meet the expectations the public has for corporations. Perhaps the tide is turning against the financialization of corporate behavior. Business executives might be emboldened by these developments and join the growing chorus of voices supporting a longer-term and more expansive view of their responsibilities and put in place the practices that ensure that they will in fact perform well in the long term. Perhaps it is time for the Business Roundtable to again review its statement on the responsibilities of the corporation to its various stakeholders.

The Next-Generation Work Force

The next-generation work force needs to be well prepared to contribute to a high-road economy and to high-road employers. It also needs to be powerful enough to incentivize more firms to move in the high-road direction. The comments from students in the online course like those reported in earlier chapters and students' priority rankings in the negotiation exercise discussed above suggest that next-generation workers are ready and willing to do this. They want to address big problems and work in flexible ways to achieve a sensible work-life balance. At the same time, they share the interest of previous generations in earning a fair and decent living, having a voice in how they work, and feeling good about the relationships they build with co-workers, managers, and the customers they serve.

These data map well with what other surveys and studies of the so-called millennial generation have found.[3] Perhaps the best summary of this generation's interests and aspirations is that they have an expanded set of expectations—they want the same things previous generations wanted from work and more.

Making these expectations a reality has to start with actions on the part of the work force itself. So let's examine what employees need to do to be productive contributors to the workplaces of the future.

Next-generation workers know that the place to start is with a good education, but in their case education has to be a more continuous process, not a phase that one

[3] For one example, see "Mind the Gaps: Deloitte Millennial Survey 2015," https://www2.deloitte.com/global/en/pages/about-deloitte/articles/millennialsurvey.html. For a more extended discussion of what millennials want from work, see Lauren Stiller Rikleen, *You Raised Us—Now Work with Us: Millennials, Career Success, and Building Strong Workplace Teams* (Chicago: American Bar Association, 2014).

completes prior to entering the work force. The mantra each generation has heard to "get as good an education as you can" is still a necessary starting point, but it is not enough. Today's students and members of the work force not only need to work hard to get the best education possible, they also need to acquire the knowledge and skills that are in high demand and continue on a path of lifelong learning throughout their careers.

But this does not mean that everyone wants to, needs to, or will be able to afford to enroll in a four-year college after completing high school. Some will prefer to pursue technical jobs and careers via vocational schools, community colleges, and/or apprenticeship programs, while others will go to four-year universities.[4] Regardless of which postsecondary education option a person chooses, it must lay a solid foundation for a career of lifelong learning.

The evidence on what to look for in providers of technical or vocational education is clear: programs that are tightly linked to industry mentors; programs that provide on-the-job apprenticeships, internships, or other learning opportunities; and programs that have well-established pathways to jobs and careers after graduation. Young people need to do their homework and choose programs that have these attributes and then vote with their feet to help get rid of programs that don't meet these standards.

PayScale, a web-based survey organization, provides additional useful information by publishing a list of the top ten college majors that are most likely to lead to underemployment after a student graduates with a bachelor's degree. Table 8.1 provides the first half of PayScale's list, including the percentage of degree holders in these fields of study who are underemployed. While we might not take these numbers or even the specific rankings too seriously, the point is clear: Choice of college major does matter. This is not a reason to avoid a major of deep interest. Young people should continue to follow their passions as undergraduates and get as broad an education as possible. The key is to come out of undergraduate school with an ability to continue learning—either by enrolling in graduate school on a full- or part-time basis or by taking advantage, perhaps with the help of employers, of the growing number of online courses various universities offer.

[4] Peter Cappelli, *Will College Pay Off?* (New York: Public Affairs, 2015).

Table 8.1. PayScale's Top Five Most Underemployed College Majors, 2016

Field	Typical Job(s)	Underemployed (%)	Reasons	
			Not Using Education or Training (%)	Part-Time, Seeking Full-Time (%)
Phys ed teaching	Phys ed teacher	56.4	79.1	20.9
Human services	Case manager; human resources associate	55.6	82.2	17.8
Illustration	Graphic designer	54.7	74.5	25.5
Criminal justice	Paralegal; fraud investigator; police officer	53.0	87.4	12.8
Project management	Project manager	52.8	91.5	8.5

Source: "The 5 Most Underemployed Bachelors' Degrees," http://www.payscale.com/career-news/2016/06/5-underemployed-bachelors-degrees.

Universities and community colleges are eager to develop markets for their online courses. This offers a second chance to those who may not have chosen undergraduate degrees that are good matches with the technical or behavioral knowledge and skills employers are looking for. Employers say they are having difficulty filling certain jobs related to technical skills, data analysis, and information technology. Young workers might put this to the test and ask their employers to come together and work with local universities to design online courses and make them available to those who are underemployed but capable of handling introductory and advanced technical and leadership courses. If an employer is unwilling to support ongoing education and reward it with career advancement opportunities, that is a good signal to look for job elsewhere. This is the first step in using exit and mobility as a source of bargaining power, an issue we turn to below.

Where Will Bargaining Power Come From?

Taking these steps will equip the next-generation work force with potentially their most important source of bargaining power: the knowledge, skills, and abilities needed to demand good-paying and fulfilling jobs. The best and most attractive employers will compete for employees with these skills and capabilities and will provide career pathways and the lifelong learning opportunities needed to keep these skills fresh. These employers will also provide opportunities to contribute to a sustainable and fulfilling company mission. But what can workers do to ensure that these opportunities are not just limited to the isolated best high-road employers? What might they do to help make sure they become the norm?

Here is where the worker apps that are emerging from within the platform economy might be used to increase worker bargaining power and signal the need for more employers to adopt high-road strategies and practices. Apps such as Glassdoor, Turkopticon, WorkIt, and SherpaShare might be a way to spread these practices until they become the norm. Entrepreneurial worker advocates need to draw on these apps and develop additional ones and other data sources that help people find good jobs and good employers and avoid employers who are stuck on the low road.

Perhaps this is one of the things young workers can call for in the next-generation unions, networks, or professional organizations they might join and/or lead. But this is only one strategy for rebuilding worker voice and bargaining power. It should not stand alone; next-generation workers should not be afraid to consider the more traditional and more visible strategies of protests or strikes. Nor should young workers stand alone; our sense is a broad cross section of the population is ready to support collective actions that are seen as based on principles of fairness, dignity, and support for high-road practices.

Remember the collective protests of Market Basket employees, managers, and customers. This was an unlikely coalition to support a strike if there ever was one. No one was unionized. Some of the managers made it clear that they are political conservatives who never before thought they would stand together in something that resembled a strike. Yet the simple and clear narrative this group communicated garnered enough broad-based community, media, and political support to win the day. The lesson workers might take away from that case is that citizens are ready to stand together when they see injustice, unfairness, or unacceptable employment conditions being imposed on their peers. When protests and collective action are supported by a clear narrative that resonates with the public, they can serve as a source of bargaining power and a wake-up call to groups that would otherwise focus only on their self-interest at the expense of others.

Educators

Educators have two primary goals and responsibilities: to teach students to be literate and informed citizens and to ensure that the work force of the future has the knowledge, skills, and abilities to compete in a global, knowledge-driven economy. Today these responsibilities begin in early childhood and extend through the full working careers of the adult labor force. Given the importance of education and skill to an economy based on knowledge and innovation, educators of all sorts and levels, from those who teach in preschools to those who teach in lifelong learning programs, need to be active contributors to the next-generation social contract.

Teachers are one of the more highly organized groups in the country. Like their labor movement colleagues, they have faced direct attacks on their dignity and collective bargaining rights in states such as Wisconsin and Ohio and could well see an escalation in such attacks in the years ahead. But more than any other group of employees, they have a ready-made set of potential allies to enlist if these attacks do escalate: parents, students, former students who remember the impact of good teachers on their development, school board members, and community groups that support good education initiatives when they see them.

If ever there was wisdom in the sports adage that "the best defense is a good offense," it is in this case. Compare two state-level experiences. Despite the political attacks on teachers' unions in Wisconsin, union-led efforts to recall Scott Walker, the governor who stripped away these rights, failed in large part because too many people in Wisconsin viewed public sector workers, including teachers, as seeking to protect their economic interests at the expense of taxpayers.[5] In contrast, as mentioned in Chapter 5, in Massachusetts, a broad coalition of school superintendents, school board members, and teachers' unions formed the Massachusetts Education Partnership in 2011 to demonstrate that there is a better way to foster school improvements by collaborating and using state-of-the-art problem-solving tools to negotiate labor contracts and solve problems on a day-to-day basis. Then, in 2016, these same groups successfully defeated a statewide referendum to expand the number of charter schools in the state with the argument that expanding charters would siphon off badly needed funds for public schools. The lesson we take from these contrasting experiences is that working in broad coalitions to support improving education opportunities and fostering innovations—even innovations that call for greater focus on holding teachers and schools accountable for meeting performance

[5] Catherine Kramer, *The Politics of Resentment* (Madison: University of Wisconsin Press, 2015).

expectations—will win broad public support and be a positive offense that will defend teachers from any escalation of attacks by politicians at the state or federal level.

Next-Generation Business and Professional Schools

If you stroll through the MIT Sloan School of Management's new building and look up on the wall, you will see the school's mission boldly displayed: "To produce principled innovative leaders who improve the world." A laudable aspiration and mission! Who could disagree? But here's what one Sloan MBA student said in a focus-group interview when she was asked what she thought: "The mission is everywhere in the marketing of Sloan but nowhere to be found in the core curriculum." Another said "What principles are we talking about?" The point that these and others are making is that for years, the core curricula of Sloan and just about all other leading business and management schools in the United States have been dominated by a narrow, technical, finance-driven model of management and of the role of the corporation in society. It reflects and reinforces the short-term shareholder-based view of the firm.

But this was not always the case in business schools. Rakesh Khurana of Harvard Business School drove this point home clearly. He describes how business schools gradually gave up on the stated mission of their founders to educate and graduate leaders who would hold business to a higher professional standard, on technical and scientific grounds and on normative grounds that recognized that business leaders have an obligation to society.[6] This original mission slowly got pushed aside as the financialization of the firm came to dominate business practices in the 1980s. Today finance departments are typically the most powerful group in business schools, just as finance departments are the most powerful group in corporations.

Business schools need to change. They may already be changing at the margins. One supportive ally in this process is the Aspen Institute. Among other things, it supports the development and dissemination of case studies that illustrate corporate social responsibility. Each year it gives a set of awards to leading business professors who do research and teaching that embody social responsibility principles and tools. This is a good thing, and it has helped introduce more of these ideas into the curriculum of a number of business schools.

There is also a growing number of elective courses in management schools and human resource departments that take on a broader perspective. One of the most popular is taught at Harvard Business School by one of its most talented teachers,

[6] Rakesh Khurana, *From Higher Aims to Hired Hands: The Social Transformation of American Business Schools and the Unfulfilled Promise of Management as a Profession* (Princeton, NJ: Princeton University Press, 2010).

Rebecca Henderson. She chose the audacious title "Reimagining Capitalism" to describe her course. In our programs at Cornell and MIT, we teach similar courses with the less catchy titles of "Managing Sustainable Businesses for People and Profits" and "Forging the Future of Work."

Students may be in the vanguard of change. Student groups at our universities and many peer institutions have begun to mobilize and question whether their courses adequately address the pressures they see building around them. A student-led group called Net Impact[7] now numbers about 50,000 and meets annually to compare notes and ideas about how to get more environmental and social sustainability content into university courses. At Harvard, students voiced their support for a strike by dining service workers. At MIT, student groups are organizing across departments to counter the rise in hateful words and actions that have surfaced recently. At Cornell, students have formed the Cornell Environmental Cooperative to mobilize student energy and a passion for creating environmental solutions for the campus community. Cornell Students against Sweatshops is part of a national effort that focuses on justice issues as they relate to labor and human rights violations worldwide. At our universities and many others, student and faculty groups are meeting to discuss how to better understand the root causes of the divides that have surfaced in the economy and society and what actions are needed to reverse direction. So while these are all still small steps, they offer some hope that students might lead the way in reforming what they are taught about the role of business and society.

Next-Generation Worker Organizations

The primary responsibility of employee advocates is to increase job opportunities and improve the conditions of work for those they represent. But taken as a whole, the leaders of today's and tomorrow's labor movement need to promote general improvements in job opportunities, working conditions, and social justice for all members of the work force. Achieving these goals will require major transformations in strategy and practice that can reverse the long-term decline in union membership, expand the base of members and coalition partners, and develop new sources of power. This in turn will entail a wide range of experimentation in order to discover the mix of networks and organizations that are capable of recruiting, mobilizing, and sustaining members; discovering and sustaining new sources of power and influence; and working collaboratively with employers, government leaders, and allied groups in civil society. In short, unions, in coalition with other progressive groups, need to once again become the economic engine for spreading high-road practices that will provide a foundation for the next-generation social contract.

[7] https://netimpact.org/.

These are strong words, and they may not sit well with those who have dealt with recalcitrant unions in the past. We have worked with unions that have been progressive and forward thinking and have driven positive change and we have worked with unions that have resisted the changes that are necessary to keep up with evolving technologies and with the evolving needs of workers and businesses. So we are not harkening back to some idealized view of past labor-management relations. The organizations that will help the next generation of workers gain a powerful voice cannot be built as a mirror image of the unions that were dominant in the mid-twentieth century and have declined in membership and influence.

It would be easy, even understandable, for leaders of today's unions to circle the wagons in defense of what may very likely be a further escalation of efforts to weaken or eradicate them. But lashing out in an angry defense of the status quo has not worked for decades and will not be a successful strategy for the future. Labor needs public support and it needs the innovative ideas and energies of advocates who are experimenting with new tools for building bargaining power. This means continuing to support and work in coalition with community groups and with the occupation- or industry-specific groups that are mobilizing outside the constraints of labor law. It is time for unions to support a broadening of the scope of labor law reforms beyond those that are limited to strengthening the ability to organize for collective bargaining as defined and provided for under the National Labor Relations Act. Perhaps it is time for a broad-based new labor manifesto that starts with the simple narrative statement that labor rights are basic human rights and that all workers should have some form of voice at work.

There is no longer a single best way to organize or a single best way to provide collective representation. Instead, the labor movement might be a broad umbrella of loosely affiliated and mutually supportive groups of women, men, minorities, immigrants, and other groups. Some of these groups may favor collective bargaining, some may favor more advisory forms of voice, and some may mobilize outside their specific workplaces, occupations, or industries. All are ready to stand in solidarity in support of campaigns for fairness and dignity at work.

Broad public support will also be a necessary condition for engaging other key stakeholders, especially business and government leaders. It is not surprising that most employers are interested in labor-management partnerships only when they are faced with powerful unions that are willing to turn to more traditional adversarial strategies if necessary. Even politicians and elected leaders too often pay lip service to labor support; once they are in office they often hold labor at bay for fear of being perceived as catering to "special interests." Broad-based support for clearly communicated proposals that support fairness and dignity is the only way to counter this behavior.

Government Leaders

We said in Chapter 7 that the primary responsibilities of government leaders with respect to work are to manage macroeconomic policies to promote job growth and employment continuity, enact and enforce employment policies that balance the needs and interests of workers and employers in ways that align with the nation's values and public interests, and promote and support innovations in employment practices that increase productivity and the number of good jobs. We outlined a comprehensive set of actions that government leaders can take to meet these responsibilities, so we need not repeat them here. Instead, the question now is how to build support for and implement the policy actions that are needed. Among other things, it will take leadership and outreach by the government's chief labor officer—the secretary of labor.

There is ample precedent for the secretary to do so. Frances Perkins had a vision for what was needed in the New Deal and was given the opportunity to pursue it. Later secretaries under both Republican administrations, such as George Shultz (the Nixon administration) and John Dunlop and William Usery (the Ford administration), and Democratic administrations, such as Arthur Goldberg (the Kennedy administration), Willard Wirtz (the Johnson administration), Ray Marshall (the Carter administration), Robert Reich (the Clinton administration), and Tom Perez (the Obama administration), were influential and visible advocates for innovations in labor and employment policies. Each actively engaged with and enlisted the advice and support of labor and business leaders to address the key employment challenges of their day.

This tradition is harder to pursue today, given the diversity of voices that would need to be engaged. We can no longer simply bring "labor" and "business" together. Today "labor" means both leaders of established unions and professional associations and leaders of working women's organizations, immigrant groups, associations of independent contractors, and others, and "business" is comprised of an equally diverse group of large and small employers, multinational firms, and local and regional business groups. No single national committee is likely to suffice or be able to reach some rarified consensus over all of the policy issues that need to be updated and transformed. Instead, what we need is an era of sustained dialogue and engagement, evidence-based experimentation, and testing at the national, sectoral, and regional levels. In short, we need a national champion and a proactive social engineer to rebuild the dialogue across the key stakeholders who must be part of the process of forging and sustaining the next-generation social contract.

But will this happen in the near future? Are we ignoring the contemporary political realities? We share the worries that appointees to key policymaking positions might favor actions that would take the nation in the opposite direction and deepen the divide and anger that has already surfaced. If we are right, then we hope the alternative approach outlined here will serve as a counterweight that is more true to the values, aspirations, and respect for each other that we believe continue to be hallmarks of American society.

An Invitation to Action

We should not and will not wait for Washington to act—positively or negatively. We need to put to work the key message of this book, that the seeds of innovation are often planted at local levels. So too with efforts to cross the divides among the key stakeholders who need to work together to build the next-generation social contract. To this end, we have launched what we call a Good Companies–Good Jobs Initiative aimed at deepening our understanding of all the issues discussed in this book. We envision that this initiative will bring innovative leaders from business, labor, government, education, and other parts of civil society together to support actions that will move our economy and workplace practices in the right direction. Our hope is to serve as a trusted convening forum that will help rebuild mutual respect across interest lines and will serve as a launchpad for further innovation. Our hope is that many others like this initiative will spring up and help lead the way to a positive, more inclusive future. You can follow our work and contribute your ideas at our website (http://iwer.mit.edu).

The public is clearly ready for a more proactive, inclusive strategy. As we noted in Chapter 1, income inequality is now recognized as a serious national (and indeed global) moral and economic problem. Wages are on the table for discussion and recent efforts to increase minimums by states, local governments, and private businesses have been well received by the public. Editorial writers of varying political persuasions are for the first time in memory writing about the need to increase the bargaining power of workers and rebuild unions. Perhaps the messages the public is sending leaders of all our stakeholder groups will create the sense of urgency needed to get them working together.

We hope so. But time is of the essence. Let's get to work.

Video Resources

The following video and exercises covering topics in this chapter can be found at http://iwer.mit.edu/speak-up-for-work/:

Instructions and materials for Using the New Social Contract *Negotiations* Exercise

Instructions and materials for Using the New Social Contract *Individual* Exercise

Index

Note: Page numbers for figures and tables are italicized.

29338374R00128

Made in the USA
Columbia, SC
22 October 2018